EUREKA!

RICHARD PLATT

EUREKA!

Great Inventors
and their Brilliant Brainwaves

RICHARD PLATT

KINGFISHER

SIR PAUL NURSE

"If you took away everything in the world that had to be invented there'd be nothing left except a lot of people getting rained on."
Tom Stoppard

We are living in a time of unprecedented discovery and invention. Scientists have now sequenced the entire human genome; animals from sheep to cats have been cloned; the Internet is changing the way we work, teach and run our businesses; and major advances in biology, chemistry and physics are transforming medicine. New technologies such as super conductors, quantum computers and nanotechnology will provide the tools for the next leap forward in scientific progress, and a continuation of human achievement that dates back to before Galileo's invention of the pendulum in 1581.

All the discoveries and inventions detailed in **Eureka!** – starting with the pendulum – have one thing in common; individuals who, in the words of the scientist Albert Szent-Györgyi, were able to see what everybody else had seen, but think what nobody else had thought.

A few discoveries have their origins in accidents – antibiotics being perhaps the most famous – but behind some others lie many years of investigation and hard work. My own discovery of a gene which sheds light on our understanding of how all living things grow and reproduce, and how cancer cells go wrong, exemplifies the typical discovery process: a great deal of thought, a great deal of work, lots of failure, and, if you're lucky, Eureka! It hits you, and you realise that you have solved the problem.

The 'eureka moment' is incredibly exciting but incredibly brief – often a source of much frustration in the laboratories where I work. The discovery itself may not always be something that will make a huge impact on society, and it can often take many years before its importance is realised. In my own field – the biology of cancer – most of the advances that are published each month are incremental but over time help to build a picture of a system or a process which ultimately points the way to greater advances.

The discoveries presented in **Eureka!** highlight where a curious and enquiring mind can lead, and illustrate why science is so important today. Without eureka moments we cannot improve the world we live in. But scientific discoveries are not made in isolation from society. The consequences of a discovery may have major moral or ethical implications. For example, new medical genetics will provide doctors with the tools and information to help in the prevention or treatment of certain diseases, but, without safety measures, that information could be misused and lead to discrimination.

In order to shape and influence how new tools and technologies are used and to ensure appropriate safety measures are in place, it is important to have a good understanding of the underlying science, and there is no better place to start than at the eureka moment – the moment of discovery.

Paul Nurse

Sir Paul Nurse, Joint Director-General of Cancer Research UK

To my inventive niece Bethan

Publishing manager: Melissa Fairley
Designer: Mark Bristow
Picture researcher: Rachael Swann
Production controller: Nancy Roberts
DTP co-ordinator: Sarah Pfitzner
Artwork archivists: Wendy Allison, Jenny Lord
Proofreaders: Sheila Clewley, Sue Lightfoot
Indexer: Sue Lightfoot

KINGFISHER

Kingfisher Publications Plc,
New Penderel House,
283–288 High Holborn,
London WC1V 7HZ
www.kingfisherpub.com

First published by
Kingfisher Publications Plc 2003
2 4 6 8 10 9 7 5 3 1
1TR/0703/TIMES/PICA(PICA)/150MA

A CIP catalogue record for this book
is available from the British Library.

ISBN 0 7534 0819 8

Printed in Malaysia

CONTENTS

INTRODUCTION

Nothing beats the thrill of a sudden scientific discovery; and we can appreciate the excitement of an engineer or inventor who solves a tough problem with a single, smart idea. We call such dramatic leaps of knowledge 'eureka moments' after the shout of Archimedes (287–212BCE). This Greek genius made a bath-time discovery which sent him scampering naked through his city shouting "Eureka!", Greek for "I've got it!"

Archimedes had been puzzling over a tricky problem. The king, Hiero II, had given a craftsman a block of precious gold to make a crown. When it was finished, the crown weighed the same as the block, but the king suspected the sly goldsmith of switching some gold for cheaper silver. Hiero told Archimedes to prove he had been cheated – or die!

Archimedes was pondering the problem as he stepped into a full bath. The bath overflowed, and he realised that the volume of water flowing out was equal to the space his body took up.

Louis Daguerre (see pages 56–57), when congratulated on his invention of photography, had a glum reply: "You must not forget that this discovery only happened after 11 years of discouraging experiments which had dampened my spirits."

! EUREKA! Archimedes could use the same idea to test the crown. He knew that if the crown was pure gold, it would take up the same space as a gold block of equal weight. But silver is lighter than gold: a block of silver the same weight as the pure gold block would be twice as big. So a crown of mixed metals would take up more room than a pure gold one. To test the crown, he lowered a block of pure gold the same weight as the crown into a bowl of water. Next, he swapped the block for the crown. If it was pure gold the water would exactly fill the bowl as before. But when he lowered in the crown, some water overflowed.

Archimedes had proved that the crown was a mixture of gold and silver. The clever test had saved his life!

Few other scientists have so much at stake, but this does not reduce their excitement when they find what they have been searching for.

Archimedes was famous for his ideas about why objects float or sink, and for working out the space that curved objects take up... and much more.

It did not take a eureka moment to invent the Walkman. Sony chief Masura Ibuka thought of it, and relied on a team of engineers to create the pocket tape player (see pages 86–87).

Eureka moments do not take the sweat out of invention. Orville Wright had one, but it still took him and his brother Wilbur another four years to build the world's first aircraft (see pages 48–49).

The stories in this book are memorable, because they make science and technology interesting, and help remind us of great discoveries. We remember them precisely because they are unusual. But only a very few scientific discoveries are made this way. Today, our world is so complex that scientists cannot always solve problems on their own. They work in teams, or combine many older inventions to make something new. So, at the end of the book, you can read how some familiar (and some exotic) ideas developed without a shout of "Eureka!" and a naked dash.

EVERYDAY LIFE AND HEALTH

Legendary eureka moments help us understand the
genius of a couple of the most famous figures in science:
Galileo and Newton. But instants of creativity have also
helped less well-known scientists make important
discoveries that affect our everyday life and health.
These people played a part in the invention of some
clever gadgets that make our lives much easier.

THE PENDULUM

Clocks first ticked and tocked some 700 years ago, but they did not keep very good time. Few were accurate enough to measure minutes, so they had just one hand, which showed the hours. Minute hands appeared only in the 17th century CE, after Italian medical student Galileo Galilei had a brilliant idea during a very boring church service.

But who was Galileo Galilei?

Born in 1564, Galilei (d. 1642) – usually referred to as Galileo – was the son of a music teacher from the town of Pisa, in Italy. After failing as a trainee monk, he went to Pisa University, where his awkward questions won him many enemies. Galileo became interested in how objects moved and fell. Later, he gained fame by using a telescope (see pages 54–55) to prove the Earth moved around the Sun. Religious leaders (who mistakenly believed the opposite) imprisoned him for this.

Galileo's eureka moment

Galileo must have seen it swinging hundreds of times before, but it was not until his first year as a medical student that he really noticed the thurible (an incense burner) in Pisa's cathedral. A monk was swinging it to spread the smoke from incense burning inside. Galileo timed the swings using his pulse – the regular heart-beat throb he could feel in his wrist. To his surprise, each swing, big or small, took exactly the same number of beats. Galileo then tried some experiments swinging pendulums – weights on different lengths of string.

! EUREKA! He found that the weight made no difference to the swing time, but the length of the pendulum did.

Galileo suggested that the swinging of a pendulum could alternately catch and release the jagged teeth of a clock's cogwheel, keeping it turning at a constant speed. His son tried to build such a pendulum clock (reconstructed here), but he did not live long enough to complete it.

There are many stories about the moment Galileo had his great pendulum idea in the cathedral. Some believe he watched a monk light a chandelier of candles, then release it. But what Pisans still know as 'Galileo's lamp' is really an incense burner.

Marking time

Doubling the length made the swings take four times as long, and halving the length cut the time to a quarter. Galileo suggested to his tutors that doctors could use a specially marked pendulum to measure a patient's pulse. Disease speeds up the pulse, so doctors routinely check it. Galileo's device, called a pulsilogium, helped them gauge it exactly.

However, it was not until Galileo was 77 years old, and totally blind, that he thought of using the pendulum's even swing to keep a clock running smoothly. His son, Vicenzo (1606–1649), sketched the pendulum clock his father described to him. After Galileo died, Vicenzo made a model – but he could not make it tick. It was a Dutch scientist, Christiaan Huygens (1629–1695), who built the world's first working pendulum clock, in 1656.

Galileo's eureka moment kept the world running on time for nearly three centuries. Pendulums swung inside most clocks until the invention of the 1929 electronic clock.

To ensure that its Big Ben bell rings on time, the huge clock above London's parliament building has a 4m pendulum. The clock's maker built into it a special mechanism to keep it running on time, even when pigeons perch on the hands.

THE THEORY OF GRAVITY

When we drop something, why does it always fall down, and not fly off in some other direction? English scientist Isaac Newton was the first to explain gravity, the force that gives objects weight, keeps our feet on the ground, and stops the Universe from flying apart. According to tradition, inspiration came to him in 1665 in the most famous eureka moment of all time.

But who was Isaac Newton?

Brought up by his grandmother, Isaac Newton (1642–1727) devised ingenious toys as a child. By day he made mouse-powered machines; by night he lifted lanterns into the sky on kites. He grew up into a proud, quarrelsome genius who made discoveries – about mathematics, movement, force and colour – that we still use today.

Cuttings from Newton's tree still grow in Kentish orchards in England; it was a fruit like this one that triggered his eureka moment.

Newton's eureka moment

One day in late summer, the scientist was deep in thought when a falling apple caught his eye: "Why does the apple never fall upwards? Or sideways?" He guessed that some invisible force must be pulling the apple towards the centre of the Earth.

EUREKA! Newton realised that the force, which he called gravity, draws together not just apples and the Earth, but all objects. It could even explain why the Earth orbits the Sun, instead of flying off into space.

Without gravity, the Moon would travel in a straight line. The huge masses of the Earth and Moon create a strong gravitational pull. This draws them together, just as the Earth's mass draws an apple towards it. The force makes the Moon constantly 'fall' towards the Earth, turning its straight-line route through space into an ellipse (a flattened circle).

Newton's famous tree was in the garden of his home at Woolsthorpe Manor, Lincolnshire, in England. According to some versions of the story, the apple struck him on the head. This detail was probably added later to make the yarn more entertaining, but the rest is true: Newton himself told several friends the story.

Taking it further

Newton thought more about gravity, and was able to explain its many puzzles in his universal theory of gravitation. This showed that the force of gravity pulls all objects together. Its strength is in proportion to the mass of the objects. But distance reduces the force; objects twice as far apart are attracted with a quarter of the gravity.

Today, researchers measure gravity with equipment such as this torsion pendulum. Moving heavy balls closer slows the swing of a pendulum inside the central tube.

THE COTTON GIN

While a guest on a cotton plantation, American science teacher Eli Whitney watched slaves struggle to clean seeds from cotton fibres. Careful observation in 1794 – and an inventive streak – led him to build a labour-saving machine to do it. His cotton gin changed the course of American history, spreading slavery throughout the South.

But who was Eli Whitney?

Eli Whitney (1765–1825) was a farmer's son. As a boy he was clever and inventive: he made a violin when aged 14, and sold nails he made in the family forge a year later. After studying law and science at Yale University, USA, Whitney went south to become a tutor. He turned to invention in 1793 when his teaching job fell through.

Whitney's eureka moment

Whitney was staying with cotton farmer Catherine Greene. The cotton that grew on her inland plantation (below) had short fibres which took a long time to separate from the seeds. Cotton grown on the coast had long fibres, and was much quicker to process. Greene explained that even when using black slaves to do the work, it was still hard for her to compete. The problem sparked Whitney's imagination. He dreamed of inventing a machine to do the job.

Eureka! The answer came in a flash. Wires set in a spinning drum could draw the fibres through a comb that was too fine for the seeds to pass through.

Building the gin

Whitney described his machine to Phineas Miller, the young manager of the estate. Intrigued, Miller offered to pay him to build the gin (engine), in exchange for half the profit. Within ten days, Whitney had made a small model – and with it one slave could do the work of ten cleaning cotton by hand. The two partners were delighted. Whitney hurried to patent his invention – making it illegal for anyone to copy it. He got his patent in 1794, and began to make cotton gins in a New Haven workshop. But the partners did not sell the machines. Instead they processed the farmers' crops, keeping some of the cleaned cotton as a fee. It seemed like a clever scheme, but the machine was so simple that anyone who had watched it operating could see immediately how it worked. Planters ignored the patent, copied the gin, and cleaned their own cotton. Within three years, the partners' business collapsed.

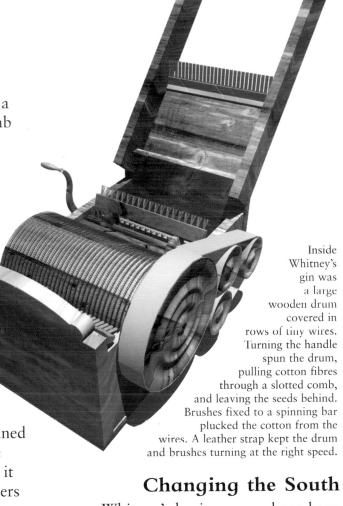

Inside Whitney's gin was a large wooden drum covered in rows of tiny wires. Turning the handle spun the drum, pulling cotton fibres through a slotted comb, and leaving the seeds behind. Brushes fixed to a spinning bar plucked the cotton from the wires. A leather strap kept the drum and brushes turning at the right speed.

Farmers still grow cotton in southern states of the USA, but they have much bigger machines than Whitney's cotton gin to help them with the harvest. Their workers are no longer slaves, and are paid for their time.

Changing the South

Whitney's business may have been a failure, but his invention certainly was not. It helped spread cotton from the coast of America's southern states to plantations far inland. Farmers who copied the gin illegally became very rich. To operate the machines – and to plant and harvest the crop – they brought thousands of slaves from Africa, often keeping them in terrible conditions. This expansion of slavery, which many Americans thought cruel and wrong, led to civil war in 1861.

In contrast to the slaves' misery, Whitney quickly recovered from his business disappointment. He won a contract to supply guns to the American government, pioneering new manufacturing methods and machines. He died a wealthy man in 1825.

VACCINATION

There is nothing small about smallpox. This deadly disease once killed a tenth of all children in Britain. Risky folk remedies helped protect a few of them, but it was a young milkmaid's bragging that led English doctor Edward Jenner to a safe remedy. Thanks to his vaccination discovery in 1788, no-one dies of smallpox today.

But who was Edward Jenner?

Edward Jenner (1749–1823), the son of a vicar, was born in the village of Berkeley, in England. From age 13 to 21, he worked as a trainee surgeon, then studied medicine in London before returning home as a doctor.

And what exactly was smallpox?

The disease started with a high fever and a bright red rash, and sometimes killed in two days. Oozing 'pocks' (blisters) scarred those lucky enough to survive longer. Smallpox was a horrible disease.

To infect eight-year-old James Phipps with cowpox, Jenner made several cuts on the boy's arm. He then rubbed in some of the pus he had taken from blisters on the hand of a patient.

Traditional cures

Edward Jenner knew of many traditional cures when he began work as a village doctor in 1773. He had heard how people protected their children by rubbing pus from a smallpox blister into skin scratches, but he also knew that this 'treatment' often passed on the disease instead of stopping it.

Jenner's eureka moment

It was milkmaid Sarah Nelmes who gave Jenner the idea of how he might control smallpox. He heard her boast that she could not get smallpox, because she once caught a much less serious disease – cowpox – from cows she milked.

EUREKA! A smallpox outbreak, in 1788, proved she was right. Jenner's patients who had caught cowpox did not get smallpox.

Testing the theory

Jenner decided to prove that cowpox protected people from smallpox. He would have to infect a healthy person with the killer disease. In 1796, Jenner plucked up the courage to try.

Injecting vaccine into a vein gives our bodies a harmless, mild dose of a disease from which we need protection. Our white cells 'remember' the infection, and react quickly and strongly when the real, dangerous smallpox germ attacks.

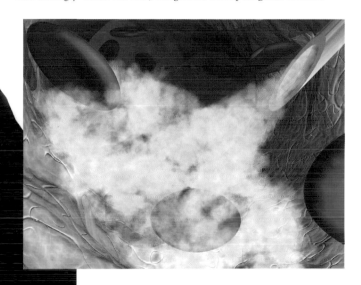

Today, Jenner's vaccination method protects us against many dangerous diseases. Instead of cowpox, medical workers inject a specially weakened form of the disease itself.

Jenner chose his gardener's son, James Phipps, and carefully infected the boy with cowpox – using pus from the blisters of a patient. Phipps got a slight fever, but soon recovered. Seven weeks later, Jenner did the experiment again – this time using pus from a smallpox sufferer. To Jenner's relief, Phipps did not get the disease. The cowpox had made the boy immune from (unable to catch) smallpox.

What happened next?

Doctors were not convinced of Jenner's treatment, which he called vaccination. Britain's Royal Society of Medicine would not report his discovery, so Jenner published a book about it. Then vaccination became popular. A new law introduced in 1845 forced everyone to be vaccinated. And by 1980, a huge vaccination scheme killed off smallpox everywhere in the world. But scientists have now suggested that terrorist organizations could reintroduce diseases such as smallpox, and these diseases may be genetically modified, and therefore vaccine resistant.

READY-MADE BUILDINGS

Putting together a building from ready-made parts makes 'instant' houses and offices possible. The first structure built like this was London's 1851 Crystal Palace. Joseph Paxton's idea for its thin steel girders and huge windows came from an unlikely source – the underside of a giant waterlily leaf.

But who was Joseph Paxton?

Son of a poor English farmer, Joseph Paxton (1803–1865), ran away from home to escape hunger and beatings. By adding a couple of years to his age, he got a job as a gardener on a rich man's estate. There, he became a skilled botanist (plant expert) and took charge of the plant glasshouses.

The monster lily

Paxton was given a tiny sample of a waterlily that grew to enormous size in its natural habitat – South America. He planted it in a heated pool inside a glasshouse. The waterlily's leaves swelled to 1.5m – big enough for Paxton's seven-year-old daughter to stand on. The lily soon outgrew the glasshouse, and Paxton wondered how he could build a new one. It needed to have lots of glass, with very few girders blocking the light.

Paxton knew nothing about the strength of glass and iron. He relied on engineers Fox, Henderson and Company to make sure that every part of the building – such as these huge central arches – would not fall down.

In the 25 weeks that it was open, the Crystal Palace attracted huge crowds. Then it was taken down, moved 12km to a new site and rebuilt – in a different shape!

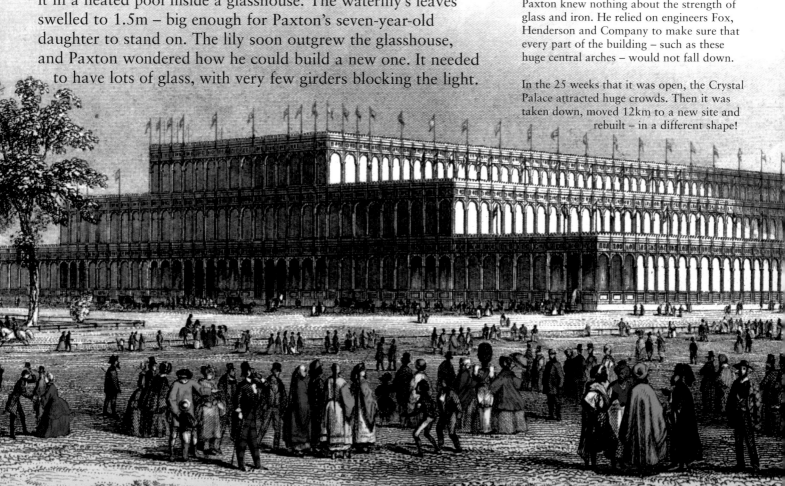

Paxton's eureka moment

The waterlily itself gave Paxton the answer. Turning over one of the leaves, he gazed at the strong ribs that spread out from the stem, which held up the leaf's surface.

! EUREKA! Paxton realised he could make a strong, lightweight building in iron and glass by copying nature.

What happened next?

Paxton built his lily house, and two years later he got the chance to build a bigger, grander version. A festival of art and industry was planned in a London park. After a committee rejected 245 designs for the exhibition hall, Paxton sent in a plan. Starting from a doodle on blotting paper, he designed the world's biggest glasshouse. By using hundreds of cast iron frames, each the same size and shape, he created a cheap, bolt-together building. Londoners loved it, and the building, nicknamed the Crystal Palace, made Paxton the world's favourite gardener.

Once the show was over, the building was moved to a new site a few kilometres away. Though fire destroyed it in 1936, Londoners still call the district where it stood 'Crystal Palace'.

The Crystal Palace was built from just a few kinds of iron parts, endlessly copied. Though made in three different factories across England, all matched, so they fitted perfectly when they reached the building site.

Today, all big buildings use some ready-made parts. This Japanese housing project goes one step further. Whole bathrooms arrive with plumbing complete, ready to bolt into position.

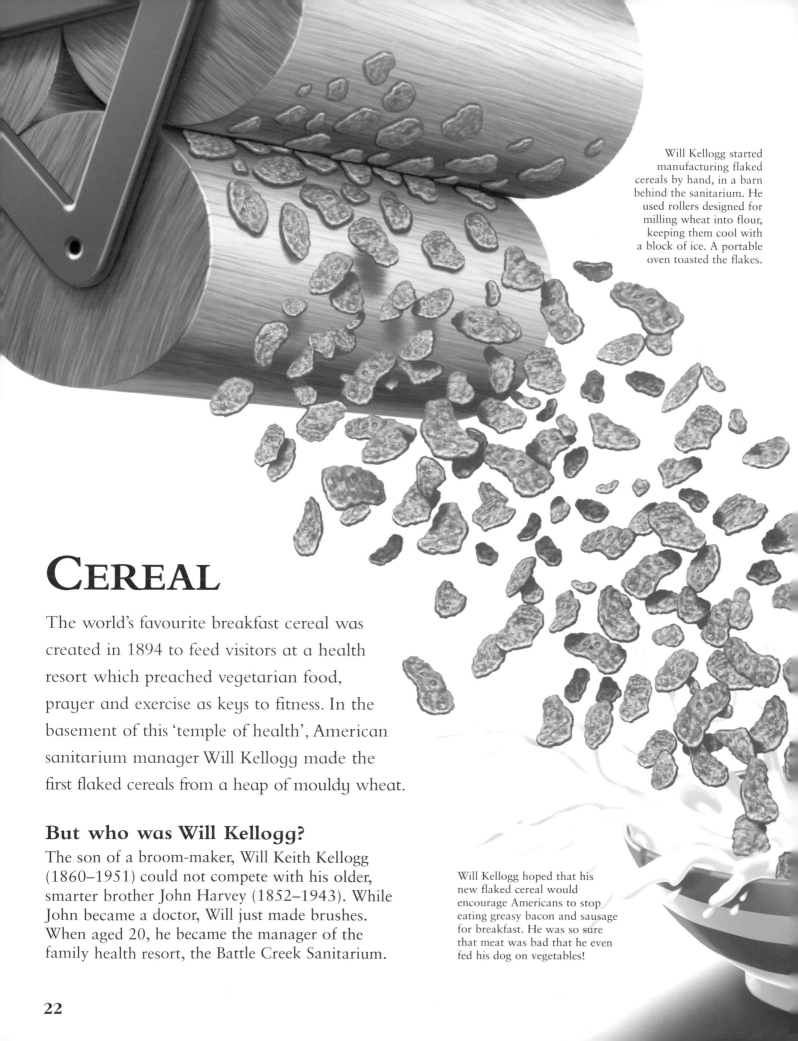

CEREAL

The world's favourite breakfast cereal was created in 1894 to feed visitors at a health resort which preached vegetarian food, prayer and exercise as keys to fitness. In the basement of this 'temple of health', American sanitarium manager Will Kellogg made the first flaked cereals from a heap of mouldy wheat.

But who was Will Kellogg?

The son of a broom-maker, Will Keith Kellogg (1860–1951) could not compete with his older, smarter brother John Harvey (1852–1943). While John became a doctor, Will just made brushes. When aged 20, he became the manager of the family health resort, the Battle Creek Sanitarium.

Will Kellogg hoped that his
new flaked cereal would
encourage Americans to stop
eating greasy bacon and sausage
for breakfast. He was so sure
that meat was bad that he even
fed his dog on vegetables!

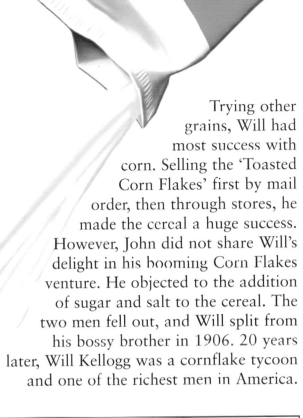

Will Kellogg's eureka moment

While John was the sanitarium's glamorous boss, Will kept the accounts and did the chores. One of his jobs was adding variety to the dull vegetarian diet. Experimenting with boiled wheat one night, Will found that it formed a sticky mess when crushed between rollers. He left most of the wheat unrolled. Returning in the morning he discovered that the wheat had gone mouldy, but decided to roll it anyway.

! EUREKA! The wheat broke up into flakes when it came out of the rollers. Will repeated the experiment, and realised that long soaking (not mould) was the key. Soaked, rolled and toasted, wheat grains made delicious flakes.

Trying other grains, Will had most success with corn. Selling the 'Toasted Corn Flakes' first by mail order, then through stores, he made the cereal a huge success. However, John did not share Will's delight in his booming Corn Flakes venture. He objected to the addition of sugar and salt to the cereal. The two men fell out, and Will split from his bossy brother in 1906. 20 years later, Will Kellogg was a cornflake tycoon and one of the richest men in America.

It was advertising that really made Corn Flakes a success. Kellogg had a flair for slogans and marketing gimmicks, promising housewives a free sample if they winked at their grocer.

"Oh ! Look Who's Here"

MADE FROM SELECTED WHITE CORN. NONE GENUINE WITHOUT THIS SIGNATURE

W. K. Kellogg

KELLOGG TOASTED CORN FLAKE CO., Battle Creek, Mich.
Canadian Trade Supplied by the Battle Creek Toasted Corn Flake Co., Ltd., London, Ont.

TOASTED CORN FLAKES

THE VACUUM CLEANER

About 100 years ago, cleaning a carpet meant lifting it from the floor, carrying it outside, and beating it until all the dust fell out. It was hard, dirty work, and engineer Hubert Booth was sure there must be an easier, cleaner way of cleaning. When he discovered there was not, he invented one in 1901 – and almost choked himself with dust!

But who was Hubert Booth?

Hubert Cecil Booth (1871–1955) was a British engineer who designed bridges, battleships and Ferris wheels (see right). But it is for vacuum cleaners, and the labour-saving contribution they made to house cleanliness, that he is now remembered.

Booth designed Ferris wheels for Blackpool (England), Paris (France), and Vienna (Austria). The first two have now been demolished, but you can still ride on his Vienna wheel, in Austria, called the Riesenrad (giant wheel). Built in 1897, it stands in the Prater amusement park, and is 64m across.

Booth's eureka moment

Booth went to watch an American inventor show off a machine that blew dust from the seats of railway carriages. It certainly worked: the jet of compressed air blew clouds of filth from the cushions – into the faces of everyone watching. Booth asked the inventor why the machine could not suck up the dirt, but he was told that this approach did not work.

Booth did not believe this, so he placed a handkerchief on a chair cushion. Pressing his mouth against it, he took a deep breath, and inhaled a lungful of choking dust! Once he had stopped coughing he turned the handkerchief over.

! ● EUREKA! On the back was a dark ring of dirt where his mouth had been. It was enough to convince him that a vacuum cleaner was practical.

Booth's monster

Booth bought an electric motor and a pump, and in 1901 built the world's first effective vacuum cleaner. There was one problem – it was too big to fit through the door of a house! This did not put his customers off. They held parties so their friends could watch the filth being sucked out of their carpets, and swept along flexible hoses to a machine in the street outside. Within seven years, other companies had brought out vacuums that were small enough to use indoors, and by 1938, two-thirds of all families had thrown away their carpet beaters and were using vacuums.

This early advertisement for a vacuum cleaner shows a housemaid using it while her mistress watches. But this was not what really happened: more often it was the mistress who operated it herself. Instead of saving women from housework, the vacuum cleaner replaced household servants.

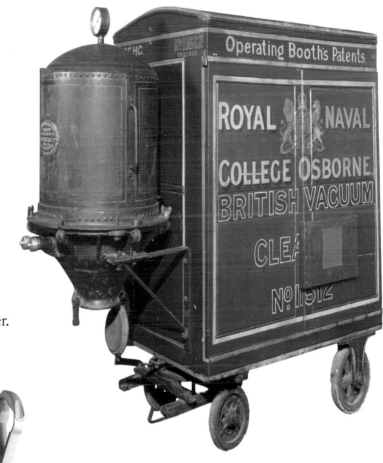

Booth did not sell vacuums. Instead he offered a cleaning service. Cleaners passed hoses through the windows, and sucked dirt from carpets into this huge machine, which was parked in the street outside. The machines were unpopular: they blocked the road, and were so noisy that passing horses bolted.

The cyclone design

After the introduction of small electric models around 1908, vacuum cleaners hardly changed for the rest of the 20th century CE. The only real improvement was the cyclone design (left). Launched in 1993, it replaced collection bags with an efficient 'whirlwind' dust separator. This did not clog as it filled so – unlike vacuums with bags – the machine kept its sucking power. Clean and slick, these bagless vacuums could not be more different from Booth's vast, noisy monster.

FROZEN FOOD

Today we expect to eat peas in the middle of winter, or prawns that once swam in an ocean on the other side of the world. Freezing and packing makes our favourite foods available everywhere, in every season. It began with Clarence Birdseye's 1912 discovery in the Arctic, where winter lasts most of the year.

But who was Clarence Birdseye?

Born in Brooklyn, in the USA, Clarence Birdseye (1886–1956) went to college to study biology, but never finished. Instead he took a job as a field naturalist with the US government, and travelled to the frozen north of Canada.

Birdseye's eureka moment

During his travels in Labrador in 1912, Birdseye watched native Americans fishing. They chipped holes in ice covering a lake. The intense cold air instantly froze the fish as it was pulled out.

EUREKA! Birdseye realised that this speedy chilling solved a problem that always spoiled frozen food: ice. Though it may seem strange, the very thing that preserved frozen food also harmed it (see main image for more details).

When food freezes slowly, ice forms in long sharp crystals. The crystals stab through the food like knives, cutting it apart from the inside. When slow-frozen food thaws, it turns to mush. But Birdseye found out that rapid freezing keeps crystals small, thus not harming the insides of the fish. So quick-frozen fish tasted as good six months later as the day it was hoisted onto the ice.

Building an Arctic factory

Birdseye realised that getting frozen food into America's kitchens was not going to be easy. He would have to find a way to stop food thawing on the way to the shop. It took him eight years to work out how to chill the food fast enough to stop the daggers of ice forming. By 1923, he had invented a machine that squeezed pre-packed food between two very cold plates, but it took him until 1930 to solve all the problems.

Unless they planned to eat frozen food the same day, families needed freezers, and few had them. In the 1930s, even refrigerators were uncommon: 2 million US homes had one, but in Britain there were only 3,000. Freezers became common kitchen fittings in the USA only in the 1950s, and much later in Europe.

Fingers of fish

In Europe, it was fish fingers that got families hooked on frozen food. They were launched in 1955 as "a new, delicious way to buy fish, which takes the time, trouble and smell out of preparing one of our favourite foods". It had taken nearly half a century, but Birdseye's brilliant idea was finally a tremendous success. Though Birdseye's name appears on millions of packs of frozen food, he was never proud of his achievement, commenting, "I do not consider myself a remarkable person. I am just a guy with a very large bump of curiosity".

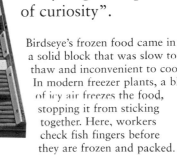

Birdseye's frozen food came in a solid block that was slow to thaw and inconvenient to cook. In modern freezer plants, a blast of icy air freezes the food, stopping it from sticking together. Here, workers check fish fingers before they are frozen and packed.

ANTIBIOTICS

Thanks to a lucky accident in 1928, Scottish doctor Alexander Fleming discovered one of medicine's most powerful weapons in the fight against germs: penicillin. However, he failed to realise its true value. The life-saving mould was turned into a drug only after Australian researcher Howard Florey stumbled on Fleming's findings in a dusty library.

Fleming's life-saving mould was similar to the many moulds that grow on old food. Magnified hundreds of times under the microscope, penicillium looks like a flower with a brush-shaped top.

But who were Fleming and Florey?

Alexander Fleming (1881–1955) studied medicine in London, in England. As an army doctor in World War I (1914–1918), he watched soldiers die from common infections. When peace returned, he searched for new antiseptic drugs.

Howard Florey (1898–1968) was a brilliant Australian who shared Fleming's interest in germ-killing drugs. By 1935, he was professor of pathology (the causes of diseases) at Oxford University, in England.

Fleming's report

Fleming studied the mould (which he called penicillium) for a while. By injecting some into a healthy rabbit, he proved that it was not harmful. But he did not take the next step – using a sick rabbit to check how good the mould was at curing disease. Instead, he simply published details of his discovery in the *British Journal of Experimental Pathology*.

Fleming's eureka moment

Alexander Fleming was a brilliant scientist, but he was messy in the laboratory. When he went away on holiday in 1928, he left behind a pile of failed experiments – jelly-filled glass dishes he used to grow bacteria. On his return, he went to wash them, and noticed that one dish stood out from the rest. A patch of mould had grown on it, and a circle of jelly around the mould was free of bacteria.

! EUREKA! Fleming realised that the mould was killing the bacteria. He thought that it might be a way of curing the diseases that bacteria cause.

Mass production of penicillin began only when it was needed to save the lives of soldiers wounded in World War II.

From mould to drug

Oxford University professor Howard Florey and his assistant Ernst Chain (1906–1979) were researching in the same field as Fleming – his article was one of 200 they read in 1939. Unlike Fleming, their research team had the know-how to extract and purify the healing agent from the mould. They tested it on eight mice, each infected with a deadly disease. Only the four treated with the (newly named) penicillin survived.

Nobel prize-winners

The lab became a penicillin factory, but the team could still not make useful quantities of the drug. Only when the United States government became interested and agreed to manufacture penicillin in quantity was Florey able to test it properly. Penicillin – the first of a group of drugs we call antibiotics – proved to be a life saver. For its development, Fleming, Florey and Chain shared the Nobel Prize for medicine in 1945.

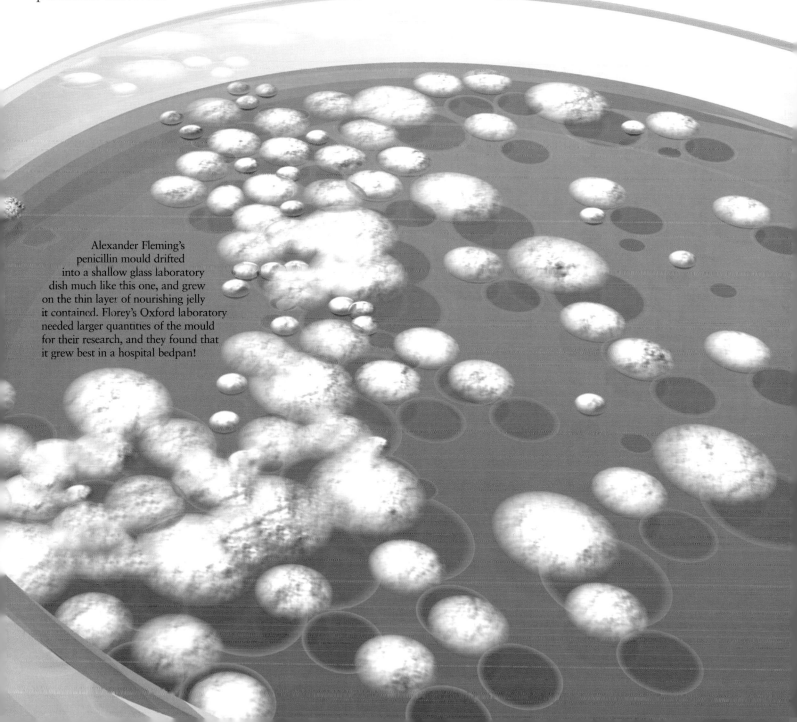

Alexander Fleming's penicillin mould drifted into a shallow glass laboratory dish much like this one, and grew on the thin layer of nourishing jelly it contained. Florey's Oxford laboratory needed larger quantities of the mould for their research, and they found that it grew best in a hospital bedpan!

NYLON

In 1930s America, a chemical company decided to invent a substitute for silk. They hired a team of brilliant young chemists to help them. Fooling around while the boss, Wallace Carothers, was away, the scientists made one of today's best-known fibres.

But who was Wallace Carothers?

Weird genius Wallace Carothers (1896–1937) was such a smart college student that he was made head of the chemistry department before he had even graduated. As a professor at Harvard University, in the USA, he studied the structure of plastics until the giant chemical company DuPont hired him in 1928.

"Get rid of the worms, Carothers!"

DuPont wanted Carothers to make a substitute for silk. This fine, costly fibre is spun by silkworms – the grubs of Chinese moths. Carothers set to work and hired eight helpers.

Years after the stretching experiment, Julian Hill acted out the moment when he drew the first threads from the sticky mass in his test-tube. It is the action of pulling and extending nylon that gives the plastic its great strength.

Under the watchful eye of their leader, the team created polymers – plastics with a chain-like structure that made them immensely strong. Their first breakthrough was *neoprene*, a kind of synthetic rubber. Just weeks later, they made another plastic, nicknamed *3-16 polymer*, that looked very promising. Carothers instructed an assistant, Julian Hill (1904–1996), to do most of the work on *3-16 polymer*.

Polymers such as nylon are created from organic chemicals – substances made by living things. Squeezing and heating the chemicals makes their structure change, from liquids and gases into strong solids. If you could see inside a polymer, you would notice that its atoms (tiniest particles) were arranged into long, very stretchy chains.

The first nylons wrinkled badly, but women did not care because they were much tougher than silk. DuPont could not make them fast enough, and shoppers fought to get the few pairs available.

The team's eureka moment

One day in 1930, while Carothers was out of the lab, Hill dipped a stirring rod into a beaker of the white, sticky plastic, and found he could pull out a thread. The thread of plastic was very stretchy and strong, so Hill and his workmates staged a tug-of-war in the corridor to see just how far they could pull it. To their amazement, stretching the threads of plastic made them stronger still.

! **EUREKA!** These plastic threads were just as springy as silk, and they could be made from oil, water and air, without the aid of moths.

What happened next?

But *3-16 polymer* turned out to be useless for weaving clothes. Ironing melted it! Carothers and his team tried for four years to find other ways to make plastic clothing fibres. When none worked, they returned to *3-16 polymer*. By tweaking the recipe in 1934, they made a handful of the 'artificial silk' that DuPont wanted. After another five years of research, factories were able to produce vast quantities of the newly named Nylon.

TEFLON

In his laboratory at DuPont, in the USA, scientist Roy Plunkett was trying to invent new cooling fluids for refrigerators and air-conditioners in 1938. He had filled a gas tank with coolant, and chilled it to an icy -79°C. But when he opened the valve the next day to release the gas, nothing came out. What happened next was to change clothes, cooking and industry forever.

But who was Roy Plunkett?

Chemist Roy Plunkett (1910–1994) joined giant American chemical company DuPont straight from college. He had been with the firm just two years when he made his lucky discovery in 1938. He worked for DuPont all his life, where his jobs included adding the lead to petrol.

Plunkett's eureka moment

When nothing came out of the tank during the cooling-fluid experiment, Plunkett and his assistant thought the tank had leaked. They put it on a scale, but it weighed the same as it did when full. The valve was not clogged, and when they unscrewed it, white waxy flakes fell out. Finally, they sawed the tank in half. Inside it was coated with the same slippery white material. This plastic-like substance had one extraordinary property: it did absolutely nothing!

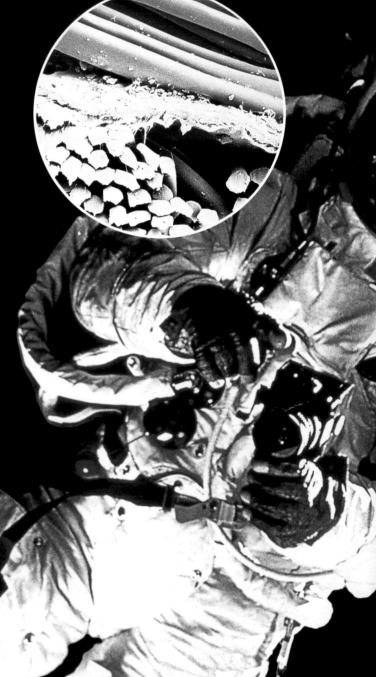

Astronauts who flew to the Moon as part of the US Apollo programme wore thick pressure suits made from 25 layers of fabric and plastic. The outer layer, made from pure Teflon fibre, stopped the suit rubbing, and protected it against fire. Two lower layers were also Teflon-coated, as shown in the close-up image.

EUREKA! Plunkett had made the dullest, slipperiest, most unreactive substance ever known. Nothing affected the new chemical.

The dullest substance on earth

The gas Plunkett had started with had the tongue-twisting name 'tetrafluoroethylene' (TFE). He guessed that the tiny gas molecules had joined up to make a long chain, called a polymer. So, Plunkett named it 'polytetrafluoroethylene' (PTFE). Later, DuPont called it Teflon. However, slipperiness is a surprisingly useful quality. Teflon could make things slip easily without oil. It could protect surfaces that were under chemical attack. One of its first uses was on the Manhattan project, America's top-secret plan to build an atomic bomb.

Non-stick pans were the idea of Frenchwoman Collette Grégoire. Her husband Marc planned to use Teflon to stop his fishing tackle tangling, but Collette suggested non-stick pans would be more useful. Calling the coating 'Tefal', they started production in 1954: this first French advert was labelled "Nothing sticks to Tefal".

To test out his invention, Plunkett plundered his laboratory cupboard. He attacked the PTFE with searing acids, powerful chemicals, and nasty liquids that dissolved almost anything. He heated it. He froze it. Still the shiny white plastic looked like new. In this recreation of one of his tests, the chemist dips a plastic rod and a Teflon rod into acid. As Plunkett found, the acid will dissolve the plastic but not affect the Teflon.

Frying pans to body parts

Peaceful uses for Teflon soon followed. When the United States began to send astronauts into space, and later to the Moon, Teflon-coated fibres protected their space suits. Back on Earth, Teflon coatings made frying pans non-stick, so cooks could fry without fat or oil. Today, there is probably some Teflon in your clothes to help garments to shrug off rain, dirt and stains. People who have heart surgery wear Teflon inside their bodies! It is the perfect stuff for 'knitting' artificial blood-vessels. Unlike other materials, it does not cause a dangerous reaction, so human tissue grows into the fabric as if it were part of the patient.

THE MICROWAVE OVEN

Puzzled by a chocolate bar melting in his pocket, ingenious American engineer Percy Spencer used the throbbing heart of a wartime radar set to pop corn and boil eggs. During World War II, he invented a whole new way of cooking, stopped his factory closing, and saved the jobs of thousands of workers.

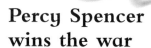

But who was Percy Spencer?

American orphan Percy Lebaron Spencer (1894–1970) never finished school, but this did not hold back his inventive nature. By the time World War II (1939–1945) broke out, he was a senior engineer with American electronics company Raytheon.

Percy Spencer wins the war

British engineers asked Raytheon to help them detect and stop bomber aircraft flying from Germany, Britain's wartime foe. The engineers had developed radar (right), which spotted the planes using microwaves – special radio waves. However, each radar set needed a device called a magnetron to generate the microwaves, and British factories could not make them fast enough.

Raytheon put Spencer on the job, and in a weekend he had figured out how to make magnetrons more quickly. Raytheon's factory was soon making magnetrons 150 times faster than they were made before.

Spencer's eureka moment

Spencer was testing a magnetron one day, when he noticed something odd. A bar of chocolate in his pocket had melted. He immediately realised that the microwaves from the magnetron had heated it. Spencer sent an assistant to buy some dry corn, and put this right in front of the magnetron. Switching on the power turned it instantly into popcorn.

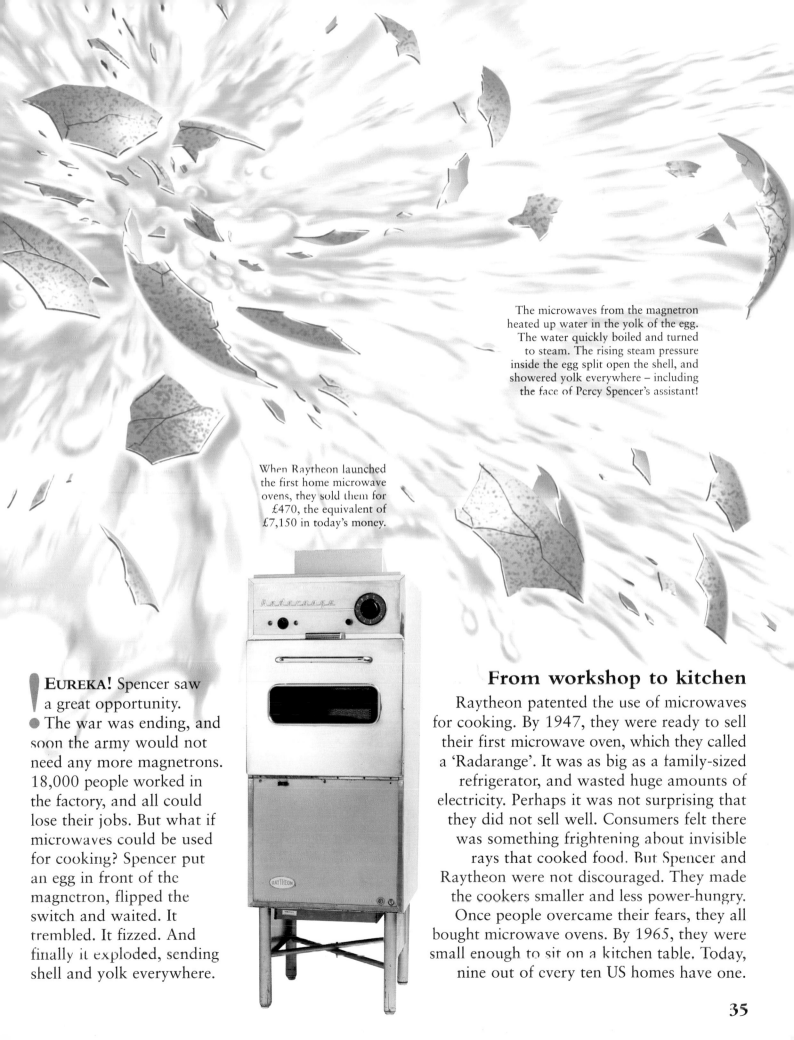

The microwaves from the magnetron heated up water in the yolk of the egg. The water quickly boiled and turned to steam. The rising steam pressure inside the egg split open the shell, and showered yolk everywhere – including the face of Percy Spencer's assistant!

When Raytheon launched the first home microwave ovens, they sold them for £470, the equivalent of £7,150 in today's money.

! EUREKA! Spencer saw a great opportunity. ● The war was ending, and soon the army would not need any more magnetrons. 18,000 people worked in the factory, and all could lose their jobs. But what if microwaves could be used for cooking? Spencer put an egg in front of the magnetron, flipped the switch and waited. It trembled. It fizzed. And finally it exploded, sending shell and yolk everywhere.

From workshop to kitchen

Raytheon patented the use of microwaves for cooking. By 1947, they were ready to sell their first microwave oven, which they called a 'Radarange'. It was as big as a family-sized refrigerator, and wasted huge amounts of electricity. Perhaps it was not surprising that they did not sell well. Consumers felt there was something frightening about invisible rays that cooked food. But Spencer and Raytheon were not discouraged. They made the cookers smaller and less power-hungry. Once people overcame their fears, they all bought microwave ovens. By 1965, they were small enough to sit on a kitchen table. Today, nine out of every ten US homes have one.

DNA FINGERPRINTING

Guilty! A killer is led away to jail, trapped by DNA fingerprinting. This powerful technology helps crime scientists identify a villain from a tiny speck of their blood, or a root of their hair. Alec Jeffreys, the British scientist who invented it, vividly remembers the autumn morning in 1984 when he made the discovery that would transform crime investigation forever.

Crime samples waiting to be analysed may hold the key to who committed the crime. And since our DNA never changes, detectives can use it to catch villains years after the crime.

But who is Alec Jeffreys?

For an eighth birthday present, Alec Jeffreys (1950–) received a chemistry set and a microscope. The gifts led to some dangerous experiments, but they also sparked a serious interest in science: he studied biochemistry and genetics at Oxford University, in England.

A stammer that identifies us all

Jeffreys went to work at Leicester University, England. There he began experiments with DNA, the long molecule that contains the genetic code (see pages 38–39). Jeffreys' team was searching for ways to spot and stop inherited diseases. These rare illnesses run in the family: people get them from their parents, and pass on to their children. Jeffreys noticed an odd quirk of DNA. In parts of it, one of the four chemical 'letters' from which the code is made was repeated over and over, like a stammer, or a stuck CD. What is more, unlike most of the genetic code, these repeated letters were always different. No two people had quite the same pattern.

Inside most of our cells is the spiralling, ladder-like DNA molecule. Pairs of chemical letters are its rungs. Their order makes DNA work like a biological computer code, describing everything about us. Half our DNA is copied from each of our parents, and the study of this transfer of DNA is called genetics.

Jeffreys' eureka moment

The importance of this discovery occurred to Jeffreys on Monday, 15 September 1984. He was in a darkroom, lifting a piece of photographic film from the developing solutions.

EUREKA! He suddenly realised that if the pattern on the film was different for everybody, it could be used to identify them. "It was so obvious…" he remembers, "…we had stumbled on a way of establishing a human's genetic identification. By the afternoon we had named our discovery DNA fingerprinting."

Jeffreys' technique snips up the 2-m-long strand of DNA into short pieces, and separates the stammers, marking them with coloured dyes. Racing the strands down a fine tube or across a sheet of jelly separates the longest ones, making tell-tale patterns that are easily matched.

Archaeologists also use DNA fingerprints to trace ancestors. By studying the DNA of this mummy, preserved in the dry, salty sand of Ürümchi, northwest China, scientists proved that it was the body of a man whose ancestors travelled from Europe 3,000 years ago.

Solving crime

Fingerprints are a valuable tool in solving crime because they too are unique – no two people have the same pattern of ridges on their fingertips. Crafty criminals know this, and wear gloves, but gloves cannot stop them leaving traces of DNA. It is hard for a suspect to deny they were involved in a crime if a sample of their DNA matches skin, hair, spittle or blood found at the crime scene. DNA fingerprinting was first used in 1985 to clear a man wrongly accused of murder. Since then it has become detectives' most important tool in solving crime.

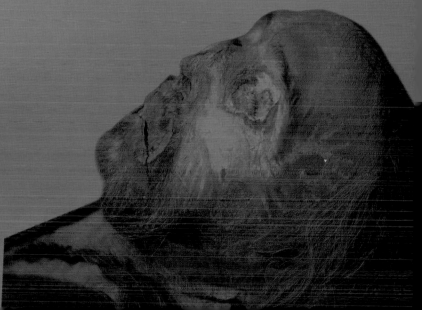

SEQUENCING THE CANCER GENE

When cancer attacks, our bodies lose control of cell growth. Cancer cells multiply, forming a deadly tumour. To find a way to control cell growth – and perhaps cure cancer – researcher Paul Nurse did not study the human body. Instead, in 1974, he looked at yeast, a tiny organism that puts the fizz in beer.

But who is Paul Nurse?

Born in England in 1949, Paul Nurse became interested in science when he was given a telescope for his eighth birthday, and used it to look at one of the first spacecraft, *Sputnik II*. Fascinated by birds and plants, he studied biological science at university, then specialized in cell biology.

The cancer gene

Like many other researchers, Nurse studied DNA. This chemical is curled inside most cells of the human body. It is shaped like a twisting ladder. Groups of 'rungs' are called genes. Together they provide a construction plan for building a human being, so DNA is sometimes called the 'genetic code'. For instance, there is a gene for brown eyes, and another for red hair. Nurse was looking for a gene that controlled cell growth. Working out exactly which gene it is – and how it works – was difficult because humans have up to 400,000 genes. In 1974, he hit upon an idea of how he could make his work easier. What if other living things relied on the same gene to control cell growth? Perhaps by studying something simpler, he could work out what happens in human cells. Nurse chose yeast. This tiny organism has a genetic code 80 times simpler than human DNA.

Yeast (magnified 30,000 times here) and people seem very different, but nearly two thirds of yeast DNA is the same as human DNA. This is because both are descended from the same ancestor, which lived some 1,000 million years ago. From this simple, one-celled living thing came a multitude of very different species (types) of plants and animals on earth today.

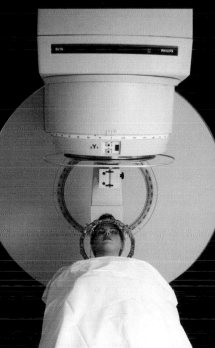

Cancer cells like this one cause harm by damaging surrounding tissues and organs (healthy cells en masse). Understanding how cancerous cells grow is important because one in every three people develops cancer. If we knew more about cancer, we might be able to prevent it by removing its causes.

Nurse's eureka moment

When Nurse started work it seemed like an odd idea. Even he admits that the chances of success were slim. Yet by the early 1980s, his team had found the gene that controlled cell growth in yeast. They called it *cdc2*. Soon they had worked out its structure. As Nurse's work continued, more and more people began to think that he could be right.

The final test came in 1987, when he had found the *cdc2* gene in human DNA. Working out its structure was slow and tedious. It produced masses of information that meant little until analysed by computer. Finally, the lab work was complete. All that remained was to let the computer crunch the numbers. As the results dribbled in, Nurse gazed in astonishment at his computer monitor.

EUREKA! It was just as he had predicted: the yeast and human genes were almost the same. He burst from his lab, and ran through the building spreading the thrilling news.

Because we know so little about cancer, present-day treatments are simple. Most use surgery, radiation (as shown here) or poisons to kill or remove the tumour. Nurse suggests that in the future these methods will seem as primitive as shaking a stopped clock to try to repair it. He hopes that his work will enable scientists to work like clockmakers, so they really understand what makes the cancer gene tick.

What happened next?

Modestly, Nurse did not call his work a breakthrough, but it is so important that in 2001 he shared a Nobel Prize – the top award in science. Nurse's research will not lead to an immediate cure for cancer, but it does provide scientists with a valuable new way of studying the disease. Despite many years of research, we still know little about cancer. Helped by Nurse's eureka moment, researchers hope to discover how cancer cells divide out of control, and stop them growing without harming the healthy organs.

GETTING AROUND

Inventors on the move have turned moments of
fortunate discovery to advantage – whether they were
dreaming of soaring like birds, or just walking behind
a plough-horse. Floating, flying and hovering all began
with a eureka moment. The one exception was the
safety lift as the inventor's eureka moment led to
a great marketing idea for his invention.

THE HOT-AIR BALLOON

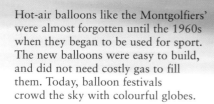

The exciting story of air travel began after a pair of French paper-makers discovered what they called 'electric smoke'. From 1783 onwards, they used it to fill vast balloons, and flew them high over the rooftops of 19th-century CE Paris, in France.

Hot-air balloons like the Montgolfiers' were almost forgotten until the 1960s when they began to be used for sport. The new balloons were easy to build, and did not need costly gas to fill them. Today, balloon festivals crowd the sky with colourful globes.

But who were the Montgolfier brothers?

The sons of a wealthy French paper-maker, Joseph (1740–1810) and Jacques (1745–1799) Montgolfier were enchanted by the idea of flying. As adults, both worked in the family paper business in Annonay, France.

The Montgolfiers' eureka moment

Joseph and Jacques figured out how they could fly when burning some rubbish in the kitchen. They noticed that flames sent scraps of paper floating up the chimney.

! EUREKA! If they could only make a craft light enough, they could use fire to lift it. They began experiments, holding ever-bigger paper bags over a fire, to send them floating upwards. They were excited to discover that each larger bag had more lifting power than the one before.

First flight

After two years of experiments the Montgolfiers created a globe of fabric, lined with paper. It was 10m across – almost as big as a circus ring. They launched it on 5 June 1783, in the town square at Annonay, near Lyon. The fire they lit underneath the balloon produced such a powerful lifting force that it took eight strong men to hold it down. When they let go, the balloon soared upwards 1,800m. The flight caused a sensation, and three months later the brothers repeated their achievement in front of their king, Louis XVI, in Paris.

Fearing that passengers would be killed, the French king at first insisted that the Montgolfiers' balloon should carry only criminals awaiting execution. But he was persuaded it was safe, and the first flight (left) carried a physician, Jean-François Pilâtre de Rozier (1754–1785), and a nobleman, François Laurent, Marquis d'Arlandes (1742–1809).

Hot-air balloons rise because heat makes air expand. This means hot air inside the balloon is lighter than cold air outside. The Montgolfier brothers did not know this. They thought the flames and sparks made a special lifting gas, which they called 'electric smoke'.

Smelly smoke

Convinced that smoke lifted their craft, they threw old shoes and rotten meat on the fire, making a stench that drove back the king. The balloon flew for eight minutes, carrying a sheep, a duck and a cockerel. Though the sheep's kicks injured the cockerel, the animals landed safely.

Soon after the Montgolfier brothers' invention, others filled balloons with hydrogen gas. These 'gas balloons' turned out to be more useful than hot-air balloons. Here (above), a soldier uses one to monitor the battlefields of France during World War I (1914–1918).

People can fly!

Success made the Montgolfier brothers bolder, and by November they had built another balloon carrying its own fire, and a much bigger basket. The new balloon rose into the skies above Paris on 21 November and carried two passengers 9km. At last humans could fly like birds!

Hot-air balloons made history again in 1999. The *Breitling Orbiter 3* used a combination of helium gas and hot air to carry its crew of two right around the world. The three-week flight won the pilots a prize of almost £610,000.

THE SAFETY LIFT

Sometimes a great invention is not enough on its own – as American Elisha Otis found out. Though his crash-proof lifts saved lives, nobody bought them. Success came only when, in a eureka moment in 1853, he thought up a way to show people just how safe his invention was. Then he could not build them quickly enough to meet public demand!

But who was Elisha Otis?

Farmer's son Elisha Graves Otis (1811–1861) drifted from job to job until he found steady work as a mechanic in a bed factory. There he devised several cunning inventions to make machines safer or better. A problem with a cargo lift in a bed factory in New York, in the USA, led to the invention that made him famous, and made skyscrapers possible.

Dangerous hoists

In 1852, bustling American factories used cargo hoists – simple lifts – to raise goods to upper floors. All suffered the same problem. If the rope broke, the lift dropped like a stone, killing anyone on board. No wonder factory workers feared them.

A new safety device

All factory workers demanded double pay to ride in the car with the cargo. So, when his employer began building a new hoist, Otis looked for ways to make the lift less dangerous. The safety device he invented was simple, but effective. It stopped the car's fall after just a few centimetres. When the bedstead firm he worked for went bankrupt, Otis set up his own hoist company. But he soon ran into trouble. Nobody was ordering his safety lifts.

The safety hoist was a strong spring fixed to the lift rope. In normal use, the lift car's weight squashed the spring flat, but if the rope broke, the spring curved down, moving two levers that forced claws out from the side of the car. The claws caught a jagged track inside the lift shaft and stopped the lift falling.

Today's tallest skyscrapers would not exist without lifts, which zoom up 60 floors in seconds. Though Otis' simple safety invention is no longer in use, lifts still have similar devices to prevent a death plunge.

Otis staged his demonstration in a tall tower called the Latting Observatory, a block north of the 1853–1854 World's Fair, in New York. At the end of the dramatic performance, Otis took off his hat and bowed as the audience applauded wildly. By repeating the stunt several times a day, Otis drew attention to the danger of lifts – and the importance of his invention.

Otis' eureka moment

While musing over the fact that hardly any safety hoists were selling, Otis figured out his brilliant advertising stunt.

EUREKA! A great exhibition was planned in New York – Otis realised it would be the perfect place to show off his invention. At the show, he set up a steam-powered lift that climbed four floors. When a crowd gathered, he jumped inside and rode to the top. There, while he explained the safety features of his new lift, a man with an axe cut the rope. Screams rose from the audience, but the car hardly dropped before the safety mechanism stopped it. Inside, Otis calmed the crowd, "All safe, gentlemen!" he shouted, "all safe".

What happened next?

Orders rolled in, and Otis built 15 hoists in 1855. Surprisingly, he did not think of hoisting people. A customer suggested this in 1856, and Otis' first passenger lift began operating the following year. Until Otis' invention, few buildings were taller than five floors, because climbing stairs any higher was exhausting. But with a lift, the sky was the limit, and the result was the skyscrapers we know today.

THE STUMP-JUMP PLOUGH

Turning forest into fields was more than most 19th-century CE Australian farmers could bear. Many left their farms after tree stumps in the soil wrecked their ploughs and injured their horses. But Richard Smith was not so easily beaten. A broken plough set him thinking...

But who was Richard Smith?

Richard Smith (1837–1919) was still a baby when his parents emigrated from London, in England, to Australia. As a young man he worked for a farm implement maker, then went into business as a blacksmith, carpenter and farmer, in South Australia.

The plough that won Smith the first prize at the 1876 Moonta show was very simple. However, it boosted wheat crops so much that people soon called the region the 'bread-basket' of Australia.

Smith's eureka moment

Smith was farming land that had only recently been cleared of trees. They had such long roots that the stumps were impossible to dig up. Each time the plough hit a stump, it jerked the horse, pulling it to a halt, or damaged the plough. This first happened to Smith when he was ploughing in 1876. His plough struck a stump, snapping a bolt that held down the plough-blade.

Though Smith's plough factory thrived and employed many workers, competitors copied his design. Fighting for credit as the plough's inventor, he apologized to a government committee for the crumpled state of his patent certificate: "I had to stow my papers in a box outside, and one rainy night a pig capsized them and made a bed of them."

To Smith's amazement, the horse did not stop. The plough's loose blade rode over the stump, and continued ploughing on the far side.

! EUREKA! Smith realised that a plough which was designed to jump over stumps would make ploughing much quicker – and cut out the tough job of digging stumps from the soil.

What happened next?

Richard and his brother Clarence (1855–1901) set to work. By June 1876, they had built a plough, and showed it at a local agricultural fair. Though it won first prize, local farmers did not like the idea of leaving stumps in the land. One shouted, "You're either fool or a lunatic, Dick Smith. A plough's got to go in hard ground and stay, unless you only want to tickle the soil".

But in the end, even the doubters bought Smith's ploughs. Before stump-jumps were invented, farmers in South Australia ploughed up just 13km² each year – an area small enough to walk around in two hours. After the invention they were turning scrub lands to field nearly 30 times faster.

The first plough Smith sold, the 'Vixen', had three blades. Each one had a separate weight to force the blade firmly back into the soil after it had jumped over a stump.

Richard Smith built a factory at Ardrossan to make his new kind of plough – and later, many other kinds of farm machine. Today, the factory is a museum and a memorial to Smith's ingenuity.

POWERED FLIGHT

'Bird men' were jokes in the early 1900s, but this did not stop two American brothers from building giant kites, and dreaming of flying in them. Orville and Wilbur Wright had trouble steering these first crude aircraft until Wilbur had a brilliant idea for controlling them.

But who were the Wright brothers?

Wilbur (1867–1912) and Orville (1871–1948) Wright ran a cycle shop in Dayton, in the USA. Neither finished school, but their father, a bishop, encouraged home study. The brothers combined scientific experiments and engineering to try to build a 'flying machine'.

Lift, control and power

The brothers began serious experiments in 1899, after reading about German glider pioneer Otto Lilienthal (1848–1896). He built and flew what we would now call hang-gliders. Inspired, the brothers began to build kites and gliders. They realised early on that to make an aircraft fly they would need to solve three problems: lift, to get it off the ground; control, so that they could go where they wanted to; and power, to move it forwards. Before long, they were building kites with plenty of lift. But control was more of a challenge. They watched pigeons in flight to see how they turned in the air, and discussed the problem for long hours when business was slack at the bike shop where they worked.

Wilbur's eureka moment

While selling a customer an inner-tube for a bike tyre, Wilbur began fiddling idly with the box. He noticed that if he held one end, and twisted the other, the normally flat surfaces of the box warped into gentle curves (see diagram above).

EUREKA! Wilbur realised that if the box was the wing of their glider, air flowing over the curved surface would make it turn.

Wing warping!

It did not take them long to put 'wing-warping' into practice on the 1.5m kite they had built. They tied strings to each corner of its wings, and launched it in a strong breeze. Making one string tighter and another looser twisted the wings. Sure enough, this steered the kite in the air – just like a modern stunt kite. They had solved the problem of control, but the brothers still had a long way to go before they could build an aircraft that would fly. Over the summer of 1900, they made a 5.2m wingspan glider model. In the autumn, they took it apart and shipped it to Kitty Hawk, in the USA, a beach with strong and steady winds. Rebuilt, the glider was strong enough to ride on. Later, they flew it as a glider. In 1901, they returned with an even bigger glider. Then, back in Dayton, they invented the wind tunnel so that they could compare the lifting ability of different wing shapes.

Flyer 1

They went back to Kitty Hawk again with an improved glider in 1902 (above), and spent the first half of the following year building a powered craft (and, incidentally, inventing the propeller). A week before Christmas that year, it flew. In an historic 59-second hop, *Flyer 1* (main image) was airborne for roughly the length of a jumbo jet. Wilbur's (now patented) wing-warp mechanism controlled its direction.

To make their 1903 aircraft fly, the Wright brothers needed an engine and propellers. When they found that all the engines they could buy were too heavy, they had a local mechanic build them a special light-weight motor. Made mainly of wood, canvas and wire, the 1903 *Flyer 1* was only just powerful enough to lift the pilot.

THE HOVERCRAFT

Looking for ways to make a craft he built go faster, boat designer Christopher Cockerell realised it was the water that slowed it down. In 1955, he figured that if he could wrap the hull in a cushion of air, it would almost skip along like a pebble spinning over the surface.

But who was Christopher Cockerell?

Brilliantly practical even as a boy, Christopher Cockerell (1910–1999) went to work for a British electronics company after leaving university. There he helped develop wartime radar sets, and patented many other inventions. When his wife inherited some money, Cockerell started a boat-building business, but soon returned to inventing.

Cockerell's eureka moment

To test out his ideas about air-cushion vehicles in 1955, Cockerell experimented with jets of air. He tried making the jet narrower and wider to see what would provide the most lifting force.

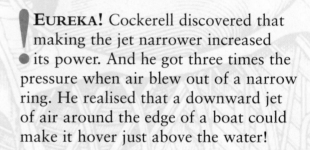

Cockerell used an industrial blower for his experiments. He pointed it down onto the scale pan of a kitchen balance. Adding weights to the other side 'weighed' the air pressure. He restricted the jet with empty coffee and cat-food tins.

EUREKA! Cockerell discovered that making the jet narrower increased its power. And he got three times the pressure when air blew out of a narrow ring. He realised that a downward jet of air around the edge of a boat could make it hover just above the water!

On its first demonstration, the hovercraft travelled from the sea to Dover's gently-sloping beach, but it was not a complete success. It threw up a huge cloud of spray, and was safe only in calm weather.

Top secret!

Cockerell took his 'hovercraft' to the British government. He told them it could travel fast over water, ice, marsh or land. Unlike a plane or ship, it did not need a runway or a harbour. Military officials liked this. However, since it did not quite belong in the army, the navy or the airforce, they declared the idea a secret... then did nothing.

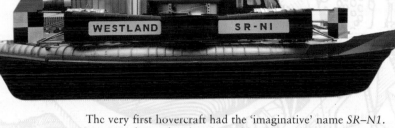

The very first hovercraft had the 'imaginative' name *SR–N1*. Here, it is shown fitted with a rubber skirt, which kept the air cushion enclosed. This greatly increased the height at which it hovered, so that small waves were no longer an obstacle.

Furious, and short of money, the inventor sold his wife's jewels to pay for hovercraft development and, in 1959, the first full-size hovercraft 'flew' across the channel between England and France.

This cross-section of a hovercraft shows how the vehicle's powerful engines draw in air and direct it downwards to provide lift. Jet engines at the rear drive it forwards. Propellors, mounted on pylons on the top, steer the craft.

The United States army experimented with military hovercraft during the Vietnam War (1954–1975), but the noise and spray they produced made them easy targets. The engines also sucked up sea-water spray, which caused corrosion (rusting) and made them unreliable.

LENSES AND LIGHT

Playing in their father's shop, a spectacle maker's
children line up lenses – and invent the telescope. This
lucky accident was just one of several that changed the
course of optics   the science of light and lenses. For
good luck and sudden bright ideas played a part in the
invention of photography and instant pictures too.

THE TELESCOPE

It may have been a children's game that gave a humble spectacle-maker the idea of lining up two lenses to magnify distant things. Hans Lippershey tried to keep this 1608 invention a secret, but news of his 'magic tube' spread quickly. In Italy, a brilliant young scientist used one to shake up our understanding of the Universe and our place in it.

To enlarge distant objects, Lippershey's young children had to choose the right lenses, and hold them the correct distance apart. It was fortunate that the combination of the two lenses worked.

But who was Hans Lippershey?

We do not know much about the inventor of the telescope. Even his name is uncertain: was it Hans Lippershey, Jan Lippersheim or Hans Lippersheim? He was born around 1570 in Wesel, in Germany, and by 1608 was living and working as a spectacle-maker in Middelburg, in the Netherlands. His shop was conveniently close to the glass factory in the town.

Lippershey's eureka moment

Hans Lippershey's two children were playing with a couple of spectacle lenses in the optician's shop. Holding them up in a straight line, and looking through both, they could see an enlarged image of storks nesting on a distant steeple.

EUREKA! Their father instantly saw the possibilities of this device, and fitted the lenses into a tube to produce what he called a 'looker'.

Lippershey's children made a lucky choice, for on their own neither of the two lenses they picked up was interesting. One was a weak magnifying glass, and the other made things look smaller.

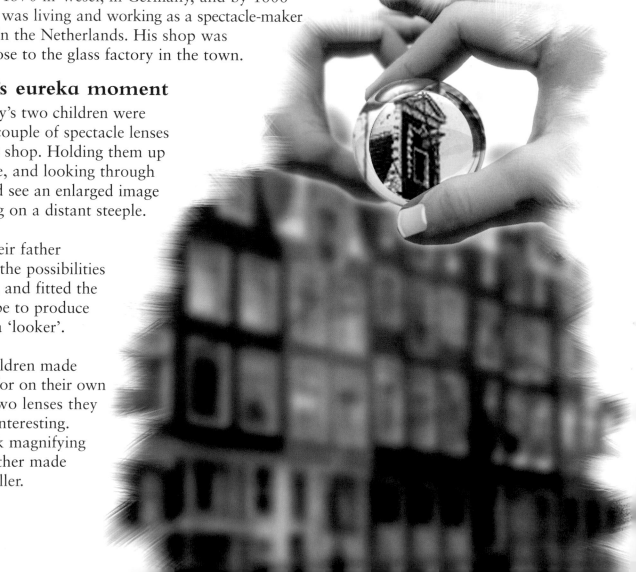

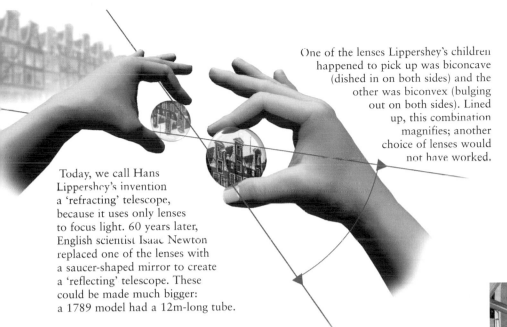

One of the lenses Lippershey's children happened to pick up was biconcave (dished in on both sides) and the other was biconvex (bulging out on both sides). Lined up, this combination magnifies; another choice of lenses would not have worked.

Today, we call Hans Lippershey's invention a 'refracting' telescope, because it uses only lenses to focus light. 60 years later, English scientist Isaac Newton replaced one of the lenses with a saucer-shaped mirror to create a 'reflecting' telescope. These could be made much bigger: a 1789 model had a 12m-long tube.

Keeping secrets

There are several other versions of this story; four centuries on, it is hard to separate truth and legend. We can be sure, though, of what followed. The spectacle-maker took his invention to the regional council, hoping to protect it with a patent. The council sent him to the Dutch Commander-in-Chief, who immediately saw how the telescope could be used in warfare. He sent the spectacle-maker away with a handsome reward, an order for three more telescopes... and a command that he should tell nobody about the invention. Despite this warning, the telescope was not a secret for long. Nine months later, Italian scientist Galileo Galilei heard a rumour about a device made from tubes and lenses that made far away things look closer.

Galileo's telescope

After just 24 hours of frantic experiments, he had produced a working telescope, and within weeks had greatly improved the instrument. Galileo's telescopes were eight times more powerful than Hans Lippershey's. They were good enough to impress the doge (duke) of Venice.

Telescopes can focus light using mirrors, which can be made bigger than lenses, creating a brighter picture.

Important discoveries

Galileo (who also devised the pendulum clock, see pages 12–13) used the telescope to make startling astronomical observations. Perhaps the most important of these were of the moons of Jupiter. By watching how they moved around the planet, Galileo was able to prove that the centre of the Universe was the Sun, not the Earth, as many people believed at the time.

The Earth's atmosphere blocks much of the light from distant stars. So, instead, some powerful telescopes such as *The Very Large Array* (below) in New Mexico, in the USA, use radio waves, focusing them with curved reflectors like television satellite dishes.

PHOTOGRAPHY

When French scene painter Louis Daguerre first showed off his newly invented photography process in 1839, the people of Paris were astonished. For each picture looked like a mirror — but a mirror that 'remembered' what it had reflected. More astonishing still was the story of how a lucky discovery had helped Daguerre to perfect the process.

New lenses that let in 16 times as much light cut the time taken to make a photograph to less than a minute. This made portraits possible – as long as the subject's head was held rigid in a clamp! Daguerre used his wife as a model for this portrait.

But who was Louis Daguerre?

Artistic and agile, Louis Daguerre (1787–1851) drew pictures and walked the tightrope with equal skill. He became a scene painter at the opera, but his ambition was to make pictures paint themselves. He tinkered with light, lenses and chemicals, and teamed up with another experimenter, Nicéphore Niepce (b. 1765). The two had not found a practical process when Niepce died in 1833.

Daguerre's eureka moment

Daguerre discovered that iodine vapour made a shiny silver plate sensitive to light. A very faint picture formed on the plate if he exposed it in a camera. Nothing made the picture less faint — until he left a plate inside a cabinet he used to store chemicals. In the morning, there was a clear image on it! A chemical in the cupboard was making the picture appear, but which chemical was it? Daguerre put a plate in the cupboard night after night, removing one chemical at a time. Even with the cupboard empty, the trick still worked. Then Daguerre noticed that the cupboard was not quite empty. In the bottom were some tiny drops of mercury from a broken thermometer.

At first, Daguerre could photograph only still objects, because even in brilliant sunlight it took many minutes to take a picture. Objects in motion during the long exposure came out as a vague blur, or vanished altogether. People thronged the streets and bridges when Daguerre made this image, but because they moved, they have disappeared, and Paris seems empty.

To develop their pictures, the first photographers put them inside a wooden cabinet. A small burner in the base heated a pool of mercury to produce the vapour that made the picture visible.

Daguerre's first photograph (above) shows plaster casts by a window. Like all 'daguerreotypes' (the inventor's name for a what we now call a photograph), it has a mirror-like finish. To see the picture, you must turn the plate to reflect something back.

EUREKA! Mercury vapour was making the images visible! Daguerre had discovered a way of developing photographs.

He planned to patent his method, but friends persuaded him not to. In return for a government pension, he published the details for all to use. His discovery was such a sensation that one famous French artist pronounced, "From today, painting is dead".

INSTANT FILM

When Edwin Land took a holiday photo of his 3-year-old daughter in 1943, her impatience sparked the American's inventive spirit. It took Land three years to satisfy his daughter's curiosity, but her simple question helped make him one of America's most admired – and richest – men.

But who was Edwin Land?

Physicist Edwin Land (1909–1991) dropped out of college to make the reflection-killing material he later called Polaroid – and because he was brighter than many of his tutors! Restlessly creative, he patented more inventions than anyone except Thomas Edison (see pages 68–69).

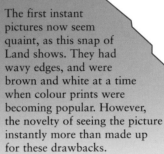

The first instant pictures now seem quaint, as this snap of Land shows. They had wavy edges, and were brown and white at a time when colour prints were becoming popular. However, the novelty of seeing the picture instantly more than made up for these drawbacks.

Land's Polaroid corporation became one of the most successful companies in the United States. Besides instant cameras, they also made weapons parts for the US army, and helped build spy planes.

Land's eureka moment

"Why can't I see it now?" Land's daughter Jennifer asked as soon as the camera shutter had clicked. "Yes…" he thought, "why shouldn't she see the photograph now?"

! EUREKA! If he could make a camera that developed the pictures, as well as taking them, everybody would want one. Because at heart, even adults are as impatient as a 3-year-old to see themselves on film.

Not all of Land's ideas were winners – 'instant movies' flopped. They needed a projector, and ran for just three minutes. They could not compete with home videos which played for an hour on a TV set.

Land joked later that he solved all the technical problems of instant photography that same afternoon "…except for the ones it took the next 19 years to solve". He meant that he had worked out the basic science in hours, but it took much longer to make a camera that worked. Land's first Polaroid camera, introduced in 1946, was not quite instant. It took a minute to make brown and white pictures. He gradually improved the process, making the pictures true black and white, and reducing developing time to just ten seconds. In 1972, he launched colour pictures that developed instantly.

You had to load two rolls of film to use Land's 1946 camera (above). After taking the picture, pulling a tab brought both rolls together, and burst a bag of chemicals between them. You peeled the film apart to see the finished picture. On Polaroid 'One-Step' film, introduced in 1972, the picture appeared while you watched.

Today's instant cameras (right) are sleek, smooth and shiny. Nevertheless there is still a thrill in watching them spit out a shiny little print that you can slip into a frame just a minute or two after you have said "cheese".

ELECTRICITY

In the ancient world people knew of electricity: the
Greeks rubbed amber to make sparks 2,600 years ago.
But it was not until the 17th century CE that scientists
began to really study electricity – with the aid of some
sparks of inspiration. Eureka moments helped bring
electric power and convenience into our homes in
the 20th century CE as well.

THE LIGHTNING CONDUCTOR

ZZZZAP! With a clap of thunder, a bolt of lightning strikes a high building. But it took a daring experiment in the 1750s, by American statesman Benjamin Franklin, to prove that lightning is a kind of electricity, and invent a protection against it.

But who was Benjamin Franklin?

Though he was also a writer and printer, Benjamin Franklin (1706–1790) is best known as an American politician. He did much to help his country break free from British rule in the 1776 revolution. He was a scientist and inventor too, and in 1752 was fascinated by newly discovered electricity.

A special kind of kite

Franklin was pretty sure that thunderstorms were electric, so he devised a simple but daring experiment. It was very dangerous so do not try and repeat it yourself! Franklin's experiment was to fly a special kind of kite in a storm. He wanted to see if electricity from the thunderstorm would flow down the kite string.

Though they protect the building they are fixed to, lightning conductors can themselves be damaged by lightning strikes. One thunderbolt was powerful enough to melt the tip of this lightning conductor, so that it flopped over limply.

This print of Franklin flying his kite with a key tied to the string made the experiment famous. However, experts argue over whether Franklin was the first, or whether he was repeating experiments of European scientists.

Safety silk

The kite had a sharp, pointed wire fixed to it to attract electricity. It also had a safety device – a silk ribbon tied to the end of the string. Electricity does not flow through silk. So, by holding the ribbon, not the string, Franklin protected himself against a shock if lightning struck. As the kite rose into the clouds, lightning struck and electricity travelled down the twine. Sparks flew from the key – just as Franklin predicted!

Franklin's eureka moment

When lightning strikes, it normally flows down to the ground through timber, brick and stone. These materials do not conduct electricity well, and the thunderbolt heats them up so much that timber burns. But if the electricity flows down through a thick metal strap, which is a good conductor, the lightning generates hardly any heat. Franklin's experiment proved that this was the case.

The Eiffel Tower in Paris, in France, bristles with lightning conductors, as it is struck by lightning several times a year.

Lightning conductors protect tall public buildings everywhere. Many of the world's tallest buildings are frequently struck by lightning. New York's Empire State Building is struck at least 100 times each year.

EUREKA! Benjamin Franklin realised that there was a simple way to stop lightning damaging tall buildings. To ensure that the thunderbolts struck the strap, and not some other part of the building, Franklin fixed a metal spike to the top. He arranged it so that the spike stuck up above the building's highest point.

Building protection

Franklin believed so strongly that lightning conductors would work that he may even have fitted them to important buildings in Philadelphia before his experiment. After his eureka moment in the storm, he was absolutely sure – and so was everyone else. Previously people had tried to protect themselves from thunderbolts using superstition. They rang church bells, or kept in their houses herbs, finches or charcoal from midsummer bonfires. None of these cures worked. But after Franklin's experiment, everyone fixed sharp rods to their roofs, and that did work. Today, every tall building is protected in this way.

THE ELECTRICAL BATTERY

While cutting up frogs on a stormy day in the late 18th century CE, Luigi Galvani noticed the legs twitched if he jabbed them with his sharp knife. Galvani thought lightning from the storm might be causing the movement.
He did not fully understand what he had discovered, but his experiments led to the electrical battery.

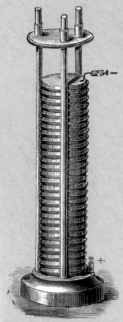

Volta found that producing more electricity was as simple as adding an extra cell – a 'sandwich' of soggy cardboard and metal discs – to his pile. Today, such an arrangement of cells, connected together, is still called a pile in the French language. In English we call it a battery.

But who was Luigi Galvani?

Italian medical student Luigi Galvani (1737–1798) became interested in anatomy and physiology – the structure of animals, and how their bodies work. It was this interest that led him, nearly 25 years later, to wire up frog legs in an electrical circuit.

Frog power

When Galvani hung frog legs on a rack outside his window to dry, he saw them twitch in thundery weather. But to his surprise, the legs twitched even on fine days.

Galvani's eureka moment

Galvani knew of Benjamin Franklin's experiments with lightning (see pages 62–63), and he was sure that electricity was causing the movement.

! **EUREKA!** Since the legs twitched when there was no thunder in the air, he thought that the animal's muscles and nerves must have been generating power.

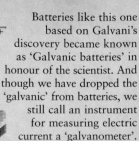

Batteries like this one based on Galvani's discovery became known as 'Galvanic batteries' in honour of the scientist. And though we have dropped the 'galvanic' from batteries, we still call an instrument for measuring electric current a 'galvanometer'.

Animal battery

He was half right. It was true that electricity was causing the muscles to shorten. But the power did not come from the frog itself. In fact, he had invented a simple electrical cell. On his drying rack, the legs touched two different metals – brass and iron. These have different abilities to attract electrons (minute particles that flow in an electric circuit). Fluid in the frog's body allowed the electrons to flow as an electric current, and it was this current that caused the muscular twitch.

Volta became famous for his battery. In 1801, he showed it to French ruler Napoleon Bonaparte (1769–1821), who made him a count and a senator. Volta's name lives on in the volt – the unit of electric force.

Galvani got it wrong

Physics professor Alessandro Volta (1745–1827) read Galvani's essay on the subject of electricity: *Commentary on the Effect of Electricity on Muscular Motion*. Volta was sure Galvani was wrong, so he tried repeating the experiments.

Building a 'pile'

Volta tried many animals and metals, and found some pairs of metals worked better than others. Finally, he stacked up zinc and silver plates, slipping in between them cardboard pads soaked in salty water. It produced far more power than a single cell (pair of plates). Between them, Galvani and Volta had invented the battery.

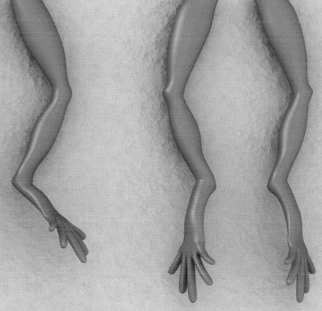

Tiny electric currents flowing in the nerves of living animals make their muscles move. Even after death, the muscles react to electricity, as Galvani discovered. But what he did not realise was that the source of the power was not the animal itself but the two metals touching it.

THE TELEPHONE

Italian–American theatre engineer Antonio Meucci thought electricity might cure migraines, so he wired a 'patient' up to a battery. But it was Meucci who got a shock, when sound travelled round his circuit! He called his 1849 invention the 'speaking telegraph'. We would call it a telephone.

Meucci made many different telephones during his experiments. He was using this one, made from turned wood, around 1858, nine years before Bell's patent. A disc of paper, stretched across the opening in the wide wooden base, moved a coil of wire inside.

But who was Antonio Meucci?

Antonio Meucci (1808–1896) emigrated from Italy to Cuba in 1835. There he set up the country's first metal coating factory, and dabbled in the fashionable trend for treating the sick with electricity.

Ouch!

Meucci put a metal plate in a migraine sufferer's mouth and connected it to a battery. Before switching the current on, Meucci put a similar plate in his own mouth, and adjusted the power to a comfortable level. Then he flipped the switch.

Bell (right) perhaps copied Meucci's invention when the two worked in a shared laboratory. Later, workers at Bell's company probably bribed officials in the US patent office to destroy documents that proved Meucci's telephone was the first. Bell grew rich from his stolen invention, and hired top lawyers to fight Meucci (left) when the inventor tried to challenge his claim.

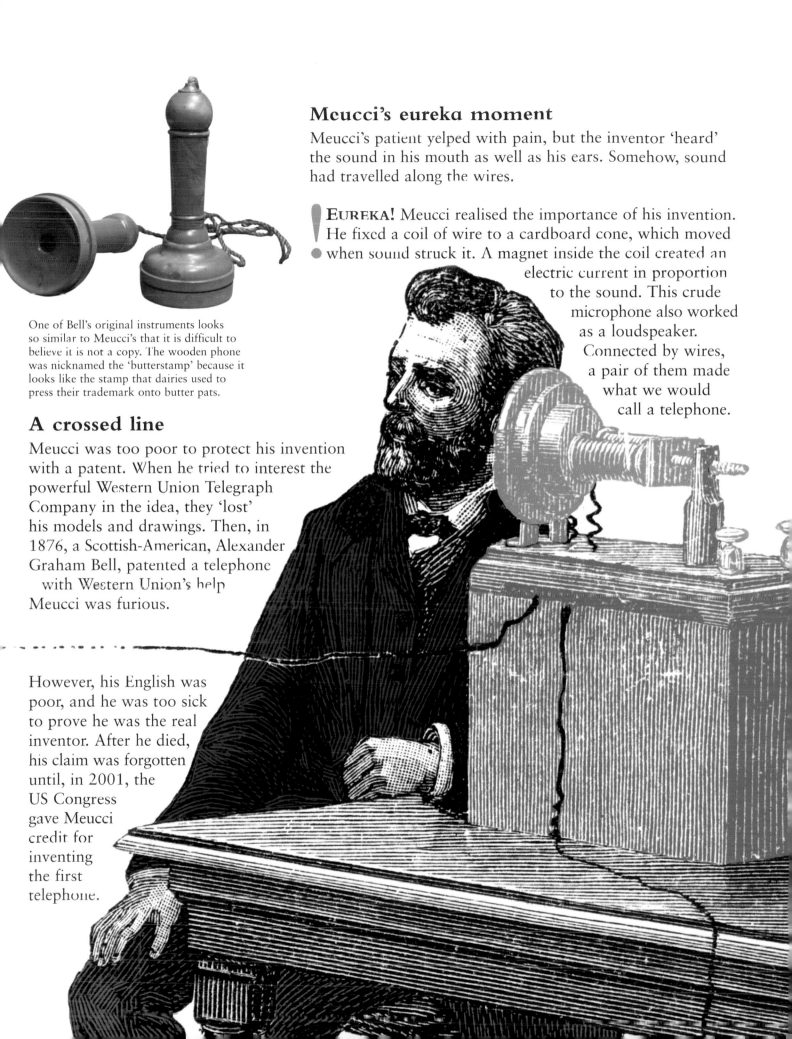

Meucci's eureka moment

Meucci's patient yelped with pain, but the inventor 'heard' the sound in his mouth as well as his ears. Somehow, sound had travelled along the wires.

! EUREKA! Meucci realised the importance of his invention. He fixed a coil of wire to a cardboard cone, which moved when sound struck it. A magnet inside the coil created an electric current in proportion to the sound. This crude microphone also worked as a loudspeaker. Connected by wires, a pair of them made what we would call a telephone.

One of Bell's original instruments looks so similar to Meucci's that it is difficult to believe it is not a copy. The wooden phone was nicknamed the 'butterstamp' because it looks like the stamp that dairies used to press their trademark onto butter pats.

A crossed line

Meucci was too poor to protect his invention with a patent. When he tried to interest the powerful Western Union Telegraph Company in the idea, they 'lost' his models and drawings. Then, in 1876, a Scottish-American, Alexander Graham Bell, patented a telephone with Western Union's help. Meucci was furious.

However, his English was poor, and he was too sick to prove he was the real inventor. After he died, his claim was forgotten until, in 2001, the US Congress gave Meucci credit for inventing the first telephone.

THE LIGHT BULB

When people called American inventor Thomas Edison a genius, he had a slick reply: "Genius is one percent inspiration… and 99 percent perspiration." He certainly sweated for months to create his 1880 light bulb, but he also had a eureka moment that helped him make it work.

When Edison's light bulb experiments dragged on longer than he expected he said, "I have not failed. I have just found 10,000 ways that will not work".

But who was Thomas Edison?

Bored by school, Thomas Alva Edison (1847–1931) stormed out when a teacher insulted him. Studying science at home, he once wired two cats together, hoping to see sparks when he stroked them. Edison got a job operating the telegraph (the forerunner of the telephone), and made a fortune inventing ways to speed up the messages it carried.

Subdividing the electric light

When Edison was a boy, electric lamps already glowed brightly – too brightly! Powerful enough for a street, they were far too bright for a room. Inventors called the challenge of lighting homes 'subdividing' the electric light – into smaller, less blinding pieces. Edison started looking for a solution in 1878. He thought it would take six weeks. However, he soon had problems with the glowing filament (metal wire) inside the glass bulb. Whatever he made it from, the bulbs burned out too soon. After a year's experiments, he had made little progress.

Edison's eureka moment

Working late one night, Edison fiddled idly with a piece of carbon cake – a mix of soot and sticky tar. Rolling it between his fingers, he made a thin sausage.

! EUREKA! Suddenly it struck him that this might work better as a filament than anything he had tried before.

Edison and his engineers developed the light bulb at a specially built laboratory in Menlo Park, New Jersey, USA. It was the first research laboratory ever created, and it is now preserved as a museum.

Until Edison developed light bulbs, the only electric lights were arc lamps which created an electric spark between a pair of pencil-sized sticks of carbon. Bright enough to light a circus tent, they were far too powerful for homes.

What happened next?

Edison relied on a team of engineers to make his good ideas work. One of them, Francis Upton (1852–1921), helped Edison turn his carbon cake roll into a filament. By October 1879, they had made a lamp that glowed for 40 hours, but it was not good enough. Nobody wanted a bulb that burned out in a week.

But Edison was not beaten. By the spring, he had built not only long lasting lamps, but wire, switches and a generator, and installed them on a ship. House and office lights soon followed. Edison's 'six week' problem had taken him two years to solve!

THE TELEPHONE EXCHANGE

The first telephones had no dials or buttons. To telephone a friend you asked an operator for the number you wanted. She used a cable to link your telephone line to your friend's. The system worked well as long as there were not many phones, and everyone knew which number they wanted. But as undertaker Almon Strowger found out in 1888, operators sometimes had their own ideas about who you wanted to call…

But who was Almon Strowger?

After studying at university, Almon Strowger (1839–1902) fought in the American Civil War (1861–1865), then worked as a schoolteacher, before setting up an undertaker's business in Kansas City, in the USA, in 1886.

Strowger's eureka moment

Strowger had become suspicious when his telephone stopped ringing. His anxiety grew when he found that another funeral parlour was doing much better than his own. On investigation, he discovered that his rival was dating the telephone operator. When people rang the exchange and asked her for an undertaker she was connecting them to her boyfriend!

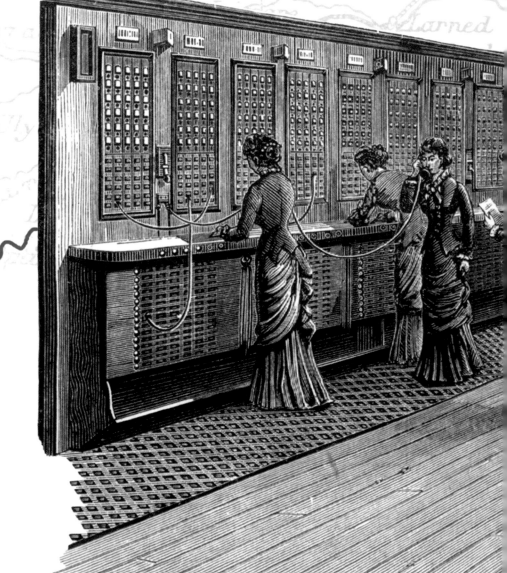

EUREKA! Strowger realised that if he devised an automatic switch, he could cut out the telephone operator. Then people who needed an undertaker could call him directly – and his business would flourish again.

With some improvements, Almon Strowger's automatic switches were installed by the million. They remained the main way of connecting calls for more than 70 years. The telephones were improved too: dials soon replaced the press-buttons, which were awkward to use, and often gave wrong numbers.

The telephone service began when women were fighting for equal rights. A telephone operator's job was 'respectable' at a time when there were few opportunities for women to work. As the telephone network grew though, no matter how many women exchanges employed, they could not keep up with the calls. Strowger's invention came just in time to prevent total chaos.

Collar-box switch

Strowger set to work, making a switch using electro-magnets, and pins pushed into a round cardboard box that once contained shirt collars. By 1891, he had devised a switch that worked, and protected it with a patent. He installed his first automatic exchange, in La Porte, in the USA, the following year.

Strowger's telephones worked with push buttons. To dial '215' you pressed the first button twice, the second button once, and the third button five times. Each press moved a switch at the exchange, connecting the call.

RADIO

Some of the world's greatest scientists studied radio waves, but it was an unknown 20-year-old Italian who sent the first radio messages. In 1894, Guglielmo Marconi started experiments with a crude transmitter, which used sparks of electricity to create radio waves. A simple receiver picked up the waves on the other side of the room, so he tried moving it much farther away...

But who was Guglielmo Marconi?

Guglielmo Marconi (1874–1937) failed the entrance exams for Bologna University, in Italy, but his father managed to sneak him in to the lectures and labs. After learning about experiments with radio, he set up a workshop at home, in an attic room once used for breeding silkworms.

Marconi's eureka moment

Marconi did not actually invent anything new. Radio waves had already been discovered; the designs of his transmitter and receiver were copied from those of other scientists. But the young Italian thought of a new use for radio waves. He realised that he could use them to send messages without wires. Nobody before him had thought of doing this.

EUREKA! Marconi sent his brother into the garden with the receiver. Alfonso (1865–1936) walked away from the house, waving a white handkerchief to show when he received a broadcast. By 1895, Alfonso carried the receiver into the next valley! His signal – a blast from a hunting rifle – proved that even a hill could not stop radio waves.

Marconi's broadcast across the Atlantic Ocean was a huge gamble. He had to build 20 antennas/aerials, each 60m high in Poldhu, UK, and Cape Cod, Newfoundland. The experiment cost £50,000 – the equivalent of nearly £3 million in today's money. Marconi worried that the earth's curvature would block the broadcast, but on 12 December 1901 his apparatus in Newfoundland received three short signals – the Morse code signal for the letter 'S'. He had done it!

Marconi received his historic signal in a station set up just below the Cabot Tower at Cape Cod, in Newfoundland. He chose this spot because nowhere in North America is closer to Europe, and because an aerial built on the high cliffs here would get much better reception than one built on low ground.

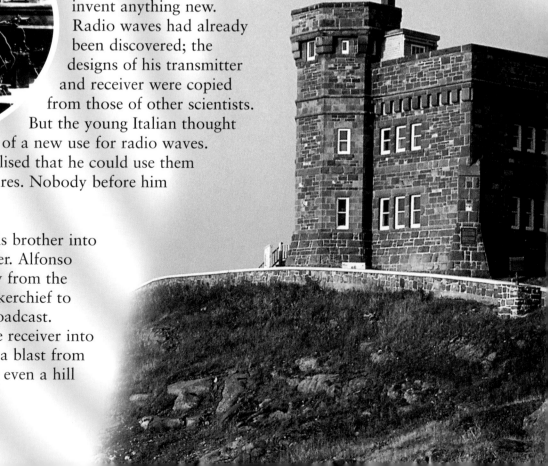

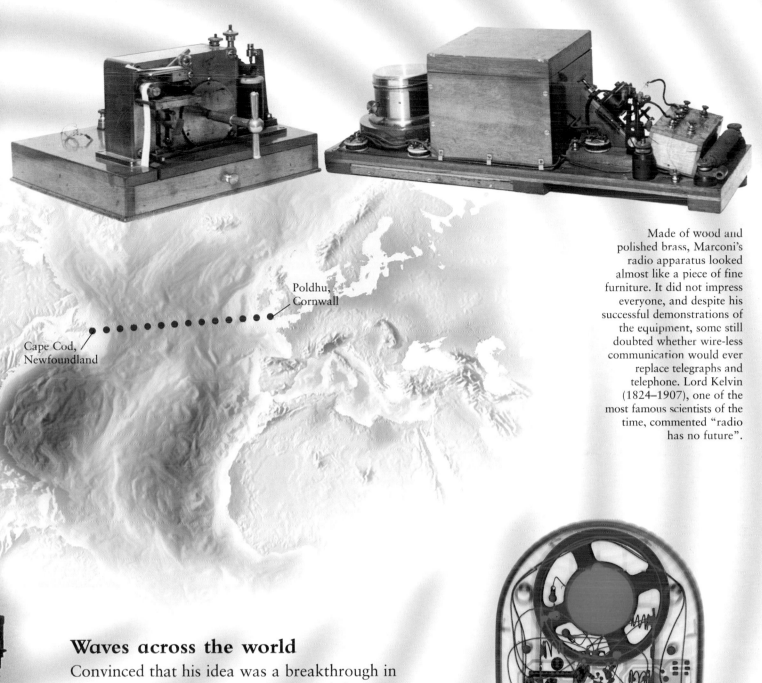

Made of wood and polished brass, Marconi's radio apparatus looked almost like a piece of fine furniture. It did not impress everyone, and despite his successful demonstrations of the equipment, some still doubted whether wire-less communication would ever replace telegraphs and telephone. Lord Kelvin (1824–1907), one of the most famous scientists of the time, commented "radio has no future".

Poldhu, Cornwall

Cape Cod, Newfoundland

Waves across the world

Convinced that his idea was a breakthrough in communication, Marconi approached the Italian Post Office. The stuffy government officials there were not interested in 'wire-less' communication. But once again, his family helped him out. His Irish-Scottish mother brought him to Britain, where a cousin introduced him to William Preece (1834–1913), the chief engineer at the British Post Office. With Preece's help, Marconi improved his apparatus so that his broadcasts reached across London, in England, and then across the English Channel. Finally, in an historic moment in 1901, Marconi broadcast a faint radio signal across the Atlantic. Suddenly the world seemed a smaller place!

Marconi made possible all our modern uses for radio, from tiny pocket receivers like this one, to portable telephones and 'cordless' computer networks.

TELEVISION

When Scottish inventor John Logie Baird gazed at the very first television pictures in 1926, he experienced an unforgettable eureka moment. Though his mechanical television system was soon overtaken by electronic competitors, it was Baird who really started the television industry. It was thanks to Baird's infectious enthusiasm that television broadcasts began – even though just 30 people had sets to watch the blurred and flickering pictures.

The first TV pictures were very poor quality because they were built up from just 30 scan lines. Though modern TV sets still scan the screen in a similar way, they use more than 600 lines, so pictures are 20 times clearer.

But who was Baird?

While his school friends were making telephones with tin cans and string, John Logie Baird (1888–1946) wired up his neighbourhood with real telephones, complete with exchange. Though his teachers called him 'slow' and 'timid', Baird managed to get a place at Glasgow University, in Scotland, to study electrical engineering.

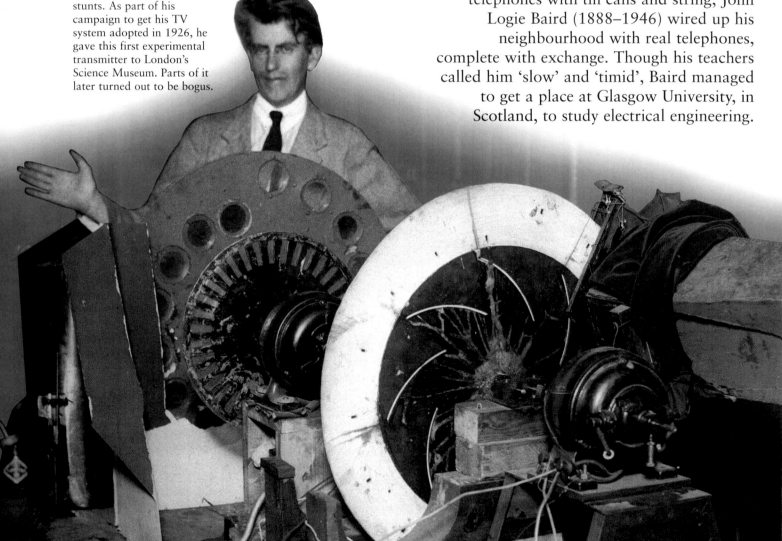

Baird had a knack for brilliant if slightly dodgy publicity stunts. As part of his campaign to get his TV system adopted in 1926, he gave this first experimental transmitter to London's Science Museum. Parts of it later turned out to be bogus.

Baird's eureka moment

Ambitious and eccentric, Baird tried his hand at making artificial diamonds and tropical fruit jams. When these ventures failed he turned to the idea of broadcasting moving pictures by radio. He raised money by day, and did research by night in a makeshift laboratory in London's seedy Soho district. His camera was a jumble of lenses, spinning cardboard discs and electric motors, all pointed at the head of a dummy called 'Stooky Bill'. Amazingly, it worked.

EUREKA! According to Baird, "The image of the dummy's head formed itself on the screen with what appeared to me an almost unbelievable clarity".

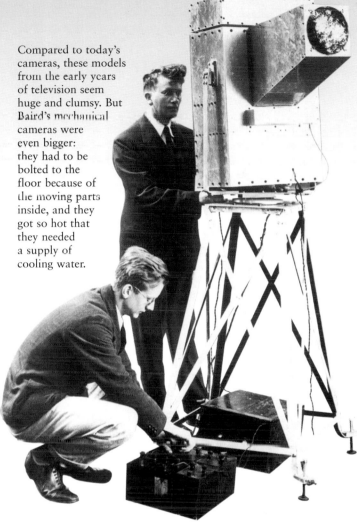

Compared to today's cameras, these models from the early years of television seem huge and clumsy. But Baird's mechanical cameras were even bigger: they had to be bolted to the floor because of the moving parts inside, and they got so hot that they needed a supply of cooling water.

A large wooden box enclosed the whirling disc that generated the picture on Baird's first TV sets. The screen, which was roughly the size of a playing card, was inside the square black tube at the top right. These receivers were so simple that some viewers built their own.

Total conviction

Not everyone agreed with Baird. Most saw only a flickering blur, but Baird was utterly convinced his system would work. Through clever publicity, determination and sheer stubbornness, he managed to get the BBC (British Broadcasting Corporation) to start experimental TV programmes.

But Baird's success did not last long. When the BBC started a full TV service in 1937, they decided that whirling discs could not produce good enough pictures. They dropped Baird's system and chose all-electronic television instead. Baird's business collapsed and he died nine years later, a bitter and disappointed man.

PATIENCE AND PLANNING

If we relied on eureka moments to inspire all science
and invention, we would live in a primitive world
without any modern conveniences. Such strokes of
genius are rare, and in these last few stories, you will
see how determination and hard work are more
reliable ways to succeed.

THE 'REAL' MCCOY

Eureka moments help us remember brilliant inventors, but for every scientist whose work made them famous, many more were overlooked. The history of technology was mainly written by white men, who did not value the contributions of female and black inventors. Most of these people are forgotten, or their work is claimed by other, more famous, inventors. Elijah McCoy is one of the few who received full credit in 1872 for his achievement.

But who was Elijah McCoy?

The son of escaped Kentucky slaves, Elijah McCoy (1843–1929) was brought up in Canada. He was a curious child and clever at fixing things, so his parents sent him to study engineering in Scotland. When he returned, the fact that he was black stopped him getting work as an engineer. He had to take a job shovelling coal and oiling locomotives.

McCoy's original automatic lubricator was like a cup of oil with a valve worked by steam. The harder the engine worked, the more the steam pressure rose – and the faster the oil flowed out. McCoy designed this improved lubricator in 1882, ten years after his first patent.

McCoy's invention

Steam engines needed constant lubrication (oiling) to keep them running. In the hours McCoy spent squirting an oil-can he wondered if he could make the train oil itself. He scraped together enough money to build a workshop, and toiled for two years to solve the problem. By 1872, he had invented an automatic lubricator, and protected it with a patent. Other engineers quickly realised the advantages of McCoy's new device, and installed it on their locos. When inferior copies became a problem, people buying lubricators asked, "Is this the 'real' McCoy?" – and the inventor's name became another word for 'quality'.

American railways were booming when McCoy was working as a stoker and part-time inventor. A railroad linked the Pacific and Atlantic coasts just three years before his first patent. Lubrication was a big problem. Engines often overheated from lack of oil, leading to delays.

SOUND RECORDING

American inventor Thomas Edison cleverly encouraged the idea that he was a creative genius. When he invented a way of recording sound in 1877, he spread a story that he then worked for four days without sleep to perfect the 'phonograph'. In reality, it was four months before he even realised the value of what he had discovered.

A distinct sound

In the summer of 1877, Thomas Edison (see pages 68–69) was trying to store and amplify signals from the newly invented telephone. He had built a device that used a needle to scratch sound waves into a moving strip of waxed paper. He shouted "Halloo" into it, and was surprised to discover that when he pulled the paper strip though the machine again he heard a distinct sound, which a strong imagination might have translated into the original "Halloo".

At the time, Edison was absorbed with his telephone amplifier, which he called a phonograph, from the Greek words for 'sound' and 'writing'. He realised much later on that he could use the machine for playing back speech. In November, he set to work to adapt his phonograph for this new purpose.

Edison's first phonograph was crude, and it needed very careful adjustment to produce any sound at all.

Edison's competitors improved on the phonograph by replacing the tinfoil with wax cylinders, which gave better sound and were more hard-wearing. The new machines, called graphophones, were widely used as office dictating machines, and to preserve historic sounds – such as the language of the Blackfoot native American people.

Phonograph sales began in 1878, sparking a craze for the machines. People queued to visit 'phonograph parlours' just to hear short pieces of music, or a few jokes by a popular comedian. However, novelty was not enough to make up for the poor sound quality and other drawbacks. After a year, the public lost interest in what Edison himself described as 'a mere toy'.

Building the recorder

Edison gave his engineer a drawing, telling him "The machine must talk". He later commented, "I did not have much faith that it would work". Neither did any of his workers – they all bet cigars on the failure of the project.

When the machine was complete, everyone gathered round. Edison wrapped a sheet of tinfoil around the cylinder, and recited "Mary had a little lamb" into the mouthpiece. After rewinding, the machine played back the nursery rhyme in Edison's distinctive squeaky voice. He said later: "I was never so taken aback in my life. I was always afraid of things that work the first time." On 6 December 1877, Edison demonstrated the phonograph at the offices of *Scientific America*.

Clerk and printer Émile Berliner (1851–1929) emigrated to the USA in 1870. He built an electrical laboratory, and his invention of a microphone won him a job with Alexander Bell's (see pages 66–67) telephone company. With his invention of gramophone records in 1887 the modern music industry began.

Discs replace cyclinders

The phonograph made Edison famous, but it was far from perfect. Recordings played for only a minute, and quickly wore out. But in 1888 hard-wearing discs replaced the cylinders, and the gramophone that played them survived almost unchanged until the introduction of CDs nearly a century later.

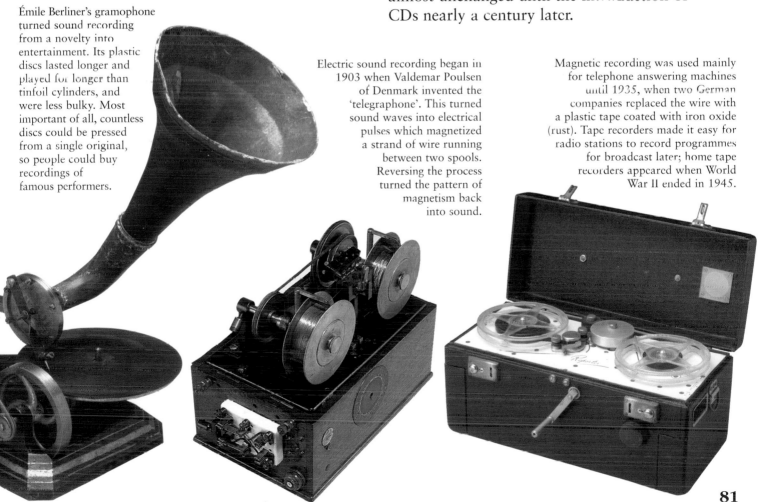

Émile Berliner's gramophone turned sound recording from a novelty into entertainment. Its plastic discs lasted longer and played for longer than tinfoil cylinders, and were less bulky. Most important of all, countless discs could be pressed from a single original, so people could buy recordings of famous performers.

Electric sound recording began in 1903 when Valdemar Poulsen of Denmark invented the 'telegraphone'. This turned sound waves into electrical pulses which magnetized a strand of wire running between two spools. Reversing the process turned the pattern of magnetism back into sound.

Magnetic recording was used mainly for telephone answering machines until 1935, when two German companies replaced the wire with a plastic tape coated with iron oxide (rust). Tape recorders made it easy for radio stations to record programmes for broadcast later; home tape recorders appeared when World War II ended in 1945.

THE SAFETY BICYCLE

More than a century ago, John Starley introduced the first modern bicycle. But it was not sudden inspiration that made his 1885 'Rover' model so popular. Starley succeeded because he brought together the spokes, tyres and chain that other inventors had perfected.

Starley's safety bicycles were safer because they had smaller wheels. Starley's first two bikes were not a great success, but his third, built in 1885, was something special. Its new diamond-shaped frame made it strong, light and safe.

But who was John Starley?

The son of a London gardener, John Kemp Starley (1854–1901) moved north to Coventry, in England, when aged 18, to work in his uncle James Starley's sewing machine business. James Starley patented the first tricycle four years later. John Starley helped him build the tricycles, but in 1877 left to start his own bicycle business.

Bicycles were not the result of a single big idea. It took a series of gradual improvements to produce a machine that was practical, safe and comfortable. In 1817, German Karl Drais (1785–1851) invented the 'running machine': a two-wheeled hobby horse (below) pushed along with the feet.

Dangerous beginnings

Scotsman Kirkpatrick Macmillan invented the bicycle in 1839. He based it on earlier 'hobby horses' which riders pushed along with their feet. However, Macmillan's new bicycle weighed almost 25kg, and was uncomfortable to ride.

Other inventors improved on Macmillan's clumsy design, but bicycles were still dangerous machines when John Starley became interested in cycling. The pedals that drove them along were fixed directly to their front wheels. To make the bicycle go fast enough, the front wheel had to be huge.

Scottish blacksmith Kirkpatrick Macmillan (1810–1878), and others, added pedals to the front wheel, creating new bicycles called 'velocipedes' (below).

Spokes, chains and tyres

John Starley's invention was the diamond frame – not the bicycle itself. However, in his Rover safety bicycle he brought together many good ideas for the first time.

Starley's 1885 bicycle did not rely just on the innovative diamond frame – the wheels also had special spokes. Spokes themselves were nothing new. They were invented 36 years earlier to make the landing wheels of a glider lighter. But Starley added a clever twist. He fixed alternate spokes to the left and right of the axle. If a wheel buckled, it could be straightened by tightening or loosening spokes. Starley made his bicycle go fast by fixing a big gear wheel to the pedals. A chain wrapped around it drove a small gear on the rear wheel. This was not Starley's idea either. Harry Lawson (1852–1925), manager of a Coventry cycle company, had made a chain-driven bicycle in 1879.

The final improvement that ensured the success of Starley's design was the air-filled tyre, introduced in 1889. Inflatable tyres, the invention of Irish surgeon John Boyd Dunlop (1840–1921), gave a much softer ride.

Setting the style

It was the special combination of improvements – and the fact that he introduced them when there was already a cycling craze – that made the Rover a winning formula. Other manufacturers copied Starley's design, and the shape of the bicycle hardly changed for a century.

Though engineers made every part of the bicycle lighter and better, Starley's basic design changed little until the 1990s. Then, new and immensely strong carbon fibre materials made possible 'one-piece' moulded frames and wheels, as on this racing bike.

With a bigger front wheel, 'high wheels' in 1868 (below) went farther – and faster – for every turn of the pedals.

Safety bicycles such as the Rover (above) had a lower, less dangerous, riding position than the high wheels that came before them. A chain drove the rear wheel.

VIDEO GAMES

The creators of the first video game did not devise it after a eureka moment. Nor did they do it for entertainment. Their aim in 1962 was to show off a brand new computer – the size of a wardrobe!

Simple beginnings

On a flickering, circular screen, a triangle and a pencil shape chase each other around a glowing blob. If you half-close your eyes, you can just about imagine they are warring spacecraft orbiting a distant planet. It is very different from the slick, realistic graphics of a modern computer console, yet in 1962 this crude, jerky game – called *SpaceWar* – amazed everyone who played it. This was the first time they had seen an interactive computer with a video screen. Other computers of the time slowly printed out the results of programs on a roll of paper, so games as we play them today were impossible.

In *SpaceWar*, to move their 'spaceships' on the round screen, and to fire missiles, players jiggled switches on the computer's cabinet. Like today's space games, rockets had limited fuel and arms and 'hyperspace', so they could disappear and reappear elsewhere.

Modern computer games look far more slick than *SpaceWar*, but the challenges are the same. Players need similar skills, too, even if the target is a lifelike monster, rather than a glowing triangle.

The DEC PDP-1 cost around £73,000 at a time when construction workers earned about 61p an hour. It was housed in a huge cabinet, and controlled by strips of paper tape punched with holes. The circular screen was the radical new feature that made the game possible.

Programming a legend

SpaceWar was the creation of Stephen Russell (1937–) and several of his friends at the Massachusetts Institute of Technology (MIT), in the USA. They wrote the game to demonstrate an amazing new computer, the PDP-1. Russell wrote the main program in a couple of months. His friends added details, including a 'death star' with lifelike gravity, and a realistic background of stars.

SpaceWar was so impressive that the computer's manufacturers, DEC, built it into every one they sold. And despite the game's simplicity, it set the standard for an entertainment industry that is now bigger than the movies.

SpaceWar was slow and clumsy partly because the computer it ran on was designed to do many other kinds of work. Modern consoles, like this PlayStation, can create realistic, fast-changing images not just because they are quicker, but also because they are built specially to run games.

THE PERSONAL STEREO

Music on the move was just a dream until the launch of the personal stereo in 1979. We may think of the tiny 'Walkman' tape players as a novel invention, yet they contained no genuinely new technology. No flash of genius led to their creation. The idea came from a businessman who wanted to relieve the boredom of long airline flights.

But who was Masura Ibuka?

The man who inspired the Sony Walkman was born in Tokyo, (1908–1997). He studied engineering at university before founding the company that would later become Sony in 1946. With his partner Akio Morita (1921–1999), he pioneered the tape recorders, televisions and miniature radios that made the Sony label famous worldwide.

One good idea and a team of inventors

As head of Sony, Ibuka spent a lot of his time flying between meetings on distant continents. To make the trips less tedious, he asked Sony engineers, in February 1979, if they could create a pocket-sized tape player, so that he could listen to high-quality stereo music without disturbing other passengers. The company's tape recorder division took just four days to build a prototype.

Before the Walkman was launched, Sony executives feared that the lack of a loudspeaker would put buyers off. At the time, headphones were big and heavy, and many Japanese people thought only the deaf wore them. So, engineers developed special headphones (left) that were so small and light that music fans would hardly know they were wearing them.

The Sony Walkman was only a little bigger than the tapes it played – most other tape players were several times larger. Sony engineers kept the size down by printing the copper connections for electronic components on both sides of the machine's circuit boards. This was a daring new idea in 1979, but today every manufacturer does it.

The creation of the personal stereo

Sony already sold journalists' miniature tape recorders, so they removed the recording mechanism, and added circuits that would make them work in stereo. The high-quality music they played delighted Masura Ibuka and Akio Morita, Sony's marketing chief. Morita was so enthusiastic that he asked the engineers to have the machines ready for sale by the start of the summer holidays, less than four months away. He also wanted the player to be cheap enough for students to buy. This dismayed the engineers. To cut the price so low, they told their boss, they would have to make 30,000 tape players – twice the monthly sales of Sony's most popular model.

A success story

The engineers raced to build the new players by the summer deadline. The new player – named Walkman – was launched in June 1979 by fashion models on rollerblades, skateboards and bicycles. There had not been time for market testing to see if the public would like the design, and for one anxious month, dealers sold hardly any of the players. But then, as word spread, customers began buying them faster than Sony could make them. The Walkman turned out to be the company's biggest success story: by 1998, Sony had sold nearly 250 million.

Sony engineers built Japan's first tape recorders in 1954, and the following year launched the first pocket-sized transistor radios. Like the Walkman, their size helped make them successful.

87

THE WEB

The Web began in the 1960s as a way to link many computers in different departments of the same company. But the Internet only became a popular success from 1976 onwards, when two programmers found a way to make it attractive and easy to use for everyone.

But who is Tim Berners Lee?

British programmer Tim Berners Lee (1955–) built his first computer while studying at Oxford University, in England. He worked as a software engineer with different companies before moving to CERN.

And who is Mark Andreessen?

While still at school, American Mark Andreessen (1971–) taught himself to program computers, and tried to write a program that would do his maths homework. As a college student he worked at the National Center for Supercomputing Applications, where he wrote Mosaic.

The web was created at the European Organization for Nuclear Research (known as CERN from the initials of its French name). Scientists there study the structure of atoms – tiny particles from which everything is made.

The birth of the web

Computers at CERN were linked by the Internet. But the information Tim Berners Lee used was scattered around many of them. So in 1976, he wrote a program to bring it all together on a single computer screen: his screen. He called the program 'Enquire Within Upon Everything', after a famous old reference book. The program used hypertext – highlighted phrases that he clicked on to jump to new 'pages' of information. In 1989, Lee realised that his idea could make it easier for anyone to use the Internet. With other CERN workers he devised a standard way of storing information, and a new language – hypertext markup language, or HTML – for writing the pages that displayed it. They named the system the World Wide Web.

The Internet grew out of ARPANET, a group of linked computers that protected the United States against nuclear missiles. Scientists began using the system in the 1970s, and today the Internet links millions of computers.

Mosaic

Lee's program worked only on computers that scientists used, until he made the code (program instructions) public. In the USA, Mark Andreessen changed it to run on ordinary PCs. He called it Mosaic. It was this 'browser' that changed a little-known network of computers into the global way of exchanging information. The growth of the Web was amazing. In 1993, there were just 50 servers (computers supplying Web pages); the following year there were 10,000. Today, there are almost 35 million servers.

Through the World Wide Web we can get pictures, sounds and words from computers scattered all over the globe. We do not need to know where the information we want is stored: we just click to find it. Today, this seems natural, but it was Lee's and Andreessen's work that made it possible.

CHRONOLOGY OF INVENTORS

LIFESPAN	INVENTOR	ORIGIN	INVENTION(S) AND PAGE NUMBER(S)
*285–211BCE	Archimedes	Greece	Principle of buoyancy, water-lifting screw 8
1564–1642	Galileo Galilei	Italy	Pendulum, motion of falling objects, motion of the planets 12
*born 1570	Hans Lippershey	Netherlands	Telescope 54
1642–1727	Isaac Newton	England	Theory of gravity, differential calculus, laws of motion, reflecting telescope 14, 55
1706–1790	Benjamin Franklin	USA	Lightning conductor, bifocal spectacles 62
1737–1798	Luigi Galvani	Italy	Battery 64
1740–1810	Joseph Montgolfier	France	Hot air balloon 42
1745–1799	Jacques Montgolfier	France	Hot air balloon 42
1745–1827	Alessandro Volta	Italy	Battery 64
1749–1823	Edward Jenner	Britain	Vaccination 18
1765–1825	Eli Whitney	USA	Cotton gin, interchangeable parts 16
1765–1833	Nicéphore Niepce	France	Photography 56
1787–1851	Louis Daguerre	France	Photography 56
1803–1865	Joseph Paxton	Britain	Ready-made buildings 20
1808–1896	Antonio Meucci	Italy/USA	Telephone 66
1811–1861	Elisha Otis	USA	Passenger lift 44
1837–1919	Richard Smith	Britain/Australia	Stump-jump plough 46
1839–1902	Almon Strowger	USA	Automatic telephone exchange 70
1843–1929	Elijah McCoy	USA	Automatic lubricator 78
1847–1931	Thomas Edison	USA	Electric light bulb, sound recording, vacuum tube 68, 80
1851–1929	Émile Berliner	Germany/USA	Sound recording on discs, microphone 81
1854–1901	John Kemp Starley	Britain	Safety bicycle 82

1860–1951	Will Keith Kellogg	USA	Corn Flakes 22
1867–1912	Wilbur Wright	USA	Powered flight 48
1869–1942	Valdemar Poulsen	Denmark	Magnetic sound recording 81
1871–1948	Orville Wright	USA	Powered flight 48
1871–1955	Hubert Cecil Booth	Britain	Vacuum cleaner 24
1874–1937	Guglielmo Marconi	Italy	Radio communication 72
1881–1955	Alexander Fleming	Britain	Penicillium 28
1886–1956	Clarence Birdseye	USA	Frozen food 26
1888–1946	John Logie Baird	Britain	Television 74
1894–1970	Percy Lebaron Spencer	USA	Microwave oven 34
1896–1937	Wallace Carothers	USA	Nylon 30
1898–1968	Howard Florey	Australia	Penicillin 28
1904–1996	Julian Hill	USA	Nylon 30
1906–1979	Ernst Chain	Germany/Britain	Penicillin 28
1908–1997	Masura Ibuka	Japan	Personal stereo 86
1909–1991	Edwin Land	USA	Instant photography, Polaroid sunglasses and filters 58
1910–1994	Roy Plunkett	USA	Teflon 32
1910–1999	Christopher Cockerell	Britain	Hovercraft 50
1921–1999	Akio Morita	Japan	Personal stereo 86
born 1937	Stephen Russell	USA	Video games 84
born 1949	Paul Nurse	Britain	Sequencing the cancer gene 38
born 1950	Alec Jeffreys	Britain	DNA fingerprinting 36
born 1955	Tim Berners Lee	Britain	World Wide Web 88
born 1971	Mark Andreesen	USA	Web browser 88

approximate lifespan

GLOSSARY

AMBER
A rare kind of hardened tree resin that looks like clear yellow plastic.

ANTISEPTIC
The ability to stop infection – or something that has this property.

ATMOSPHERE
The layer of gases, including life-giving oxygen, that surrounds our planet.

ATOMS
The tiny particles from which all matter – solid, liquid or gas – is made.

BACTERIA
Minute creatures, visible only with a microscope, some of which cause disease in plants and animals.

BROWSER
A program that enables a computer user to explore the World Wide Web and view linked words, pictures, sound and movies.

CARPET BEATER
A cane or wire paddle that was once used to beat dirt from carpets and rugs.

CIRCUIT BOARD
The plastic panel at the heart of most electronic devices. Copper wires printed on the board connect up the electronic components fixed to it.

COGWHEEL
A wheel with cogs (serrated teeth) around its edge, which enable the wheel to engage with a similar wheel, so that both turn together.

COWPOX
A mild skin disease of cattle that is sometimes caught by people who care for the animals.

ELECTRO–MAGNET
A coil of wire that becomes magnetic when electricity flows through it.

EUREKA MOMENT
A sudden brilliant idea that leads to a new discovery or invention.

FILAMENT
A thread-like part of an object, such as the thin wire of an electric light bulb that glows white-hot when electricity flows through it.

GEAR
A cogwheel, or a collection of linked cogwheels.

GENERATOR
A machine that produces electric power when operated.

GENES
Small parts of DNA, the long molecule that contains the 'design' for all living things. Passed on from parents to their children, individual genes carry the 'code' for features such as hair and eye colour.

GLASSHOUSE
A glass-walled building which uses the sun's heat to help plants grow more quickly.

IMMUNE
Protected against infection by a particular disease.

IODINE
A dark grey crystal that is antiseptic in small quantities, but poisonous in large doses, iodine combines with silver metal to make a chemical that darkens in daylight.

IRON OXIDE
A chemical with the common name 'rust' that forms when oxygen in the air reacts with wet iron and steel.

MAGNETRON
An electronic device used to produce microwave energy in an oven or radar apparatus.

MASS
How much material an object contains: the greater an object's mass, the stronger is gravity's pull upon it, and thus the greater its weight.

MERCURY
A poisonous silver metal that is liquid at room temperature.

MIGRAINE
A severe headache that often comes with sickness and blurred vision.

MORSE CODE
A series of long and short bursts of electricity, sound or light, used to carry messages.

MOULD
A type of tiny fungus that forms a fur-like coating on rotting animal or plant material.

NOBEL PRIZE
A valuable prize awarded each year for the best in physics, chemistry, medicine, literature, economics or peace.

PATENT
A special kind of protection for an invention that makes it public knowledge but stops anyone copying it until the inventor has had the chance to profit from it.

PLASTIC
A material, usually made from oil, that can be easily moulded and shaped.

POLYMER
A kind of plastic created by making many small molecules of a gas or liquid join together to form a long chain with quite different properties.

PULSE
The throbbing beat that can be felt on the body of a living human or animal as the heart pumps blood.

RADIATION
Any of a series of waves, including light, microwaves, radio and radar; but most often used to mean powerful waves like X-rays that can harm living material.

RADIO WAVE
Form of radiation commonly used to broadcast messages and entertainment.

RIB
Thin, curved chest bone – or an object that resembles such a bone.

SANITARIUM
A resort or treatment centre for those who are sick, or imagine they are.

SILK
The fine thread produced by silkworms, or the valuable fabric woven from such threads.

SILKWORMS
The caterpillar of an Asian moth, bred and raised for the silk it spins.

SLAVERY
A system in which captive people are bought and sold as possessions, and made to work, often without pay, and without the rights that free people enjoy.

SLOGANS
A short, snappy, easily-remembered phrase to advertise a product.

SMALLPOX
A dangerous, easily spread disease that kills many of those it infects, and leaves survivors' skin marked by scars.

SPOKES
Thick straight wires joining the rim of a wheel to the central hub around which it spins.

TECHNOLOGY
Applying science in a useful way.

TRANSMITTER
An instrument that sends out a communication signal.

TUMOUR
An uncontrolled growth of cells, such as cancer, that sometimes leads to ill health or death.

UNDERTAKER
Someone whose job it is to prepare the dead for burial or cremation (burning).

VACCINE
A deliberately-weakened form of a disease that, when injected or swallowed, gives immunity to infection.

VAPOUR
A gas-like form of a substance that is usually liquid or solid.

VEGETARIAN
Someone who eats no meat, or who eats only vegetables.

YEAST
A simple, microscopic plant valued for its ability to turn sugar into alcohol and carbon dioxide gas.

INDEX

ACKNOWLEDGEMENTS

The publisher would like to thank the following for permission to reproduce their material. Every care has been taken to trace copyright holders. However, if there have been unintentional omissions or failure to trace copyright holders, we apologize and will, if informed, endeavour to make corrections in any future edition.

Key: B = bottom, C = centre, L = left, R = right, T = top

5TR Cancer Research (UK); 8TC Popperfoto; 9CC Bettmann/Corbis; 9TC Bettmann/Corbis; 9RC Sony Corporation; 12CL Science Museum/Science & Society library; 12TR Science Museum/Science & Society Library; 13BR Digitalvision Limit; 14BL Getty Images/World Perspectives; 14TR Crown Copyright/Brogdale Horticultural Trust; 15BLC Courtesy of Chip Simon; 16CLC Bettmann/Corbis; 16BC Bettmann/Corbis; 17BLC Richard Hamilton Smith/Corbis; 18TRC Hulton-Deutsch Collection/Corbis; 18BLC Bettmann/Corbis; 19TR BSIP, Villareal/Science Photo Library;19BL Roger Harris/Science Photo Library; 20–21B Courtesy of Richard Platt; 20CL Science Museum, London/Heritage-Images; 20TC Courtesy of Richard Platt; 21CRC Paul Almasy/Corbis; 22BR The Advertising Archive; 24TR Getty Images; 24BL Science Museum/Science & Society Picture Library; 25TR Science Museum/Science & Society Picture Library; 25BC Dyson Graphics; 27B Rosenfeld Images Ltd/Science Photo Library; 27TR BSIP Bernard/Science Photo Library; 27C Electric Corp./National Geographic Image Collection; 28CL Hulton-Deutsch Collection/Corbis; 28C Science Photo Library; 28TRC Science Pictures Limited/Corbis; 28BRC Bettmann/Corbis; 30TR Courtesy of Du Pont; 31BL Courtesy of Du Pont; 31TRC Hulton-Deutsch Collection/Corbis; 32C Dr. Jeremy Burgess/Science Photo Library; 32CL Courtesy of Du Pont; 32 B NASA/Science Photo Library; 33TR Courtesy of the Hagley Museum and Library; 34TL Raytheon Company Photo; 34BC © Bettmann/Corbis; 35BC Raytheon Company Photo; 36CL David Parker/Science Photo Library; 36TR J. C. Revy/Science Photo Library; 37C Tek Image/Science Photo Library; 37BRC Corbis Sygma; 38B Dr Gopal Murti/Science Photo Library; 39TL Alfred Pasieka/Science Photo Library; 39CR Philippe Plailly/Eurelios/Science Photo Library; 39BC Roger Ressmeyer/Corbis; 42TC © Duomo/Corbis; 42CL Mary Evans Picture Library; 42CB Mary Evans Picture Library; 42C Sheila Terry/Science Photo Library; 42–43C © Royalty-free/Corbis; 43TC, TR, CR Vince Streano/Corbis; 43BC Popperfoto; 43TRC Corbis; 44CLC Bettmann/Corbis; 44–45C Science Photo Library; 45TRC David Lees/Corbis; 46TR Courtesy of Mortlock Library of South Australia; 47BRC Beryl E. Neumann, for the National Trust of Australia/The Smith Brothers and the Stump Jump Plough; 47TRC Beryl E. Neumann, for the National Trust of Australia/The Smith Brothers and the Stump Jump Plough; 48–49BC Bettmann/Corbis; 48CLC Bettmann/Corbis; 50BLC Hulton-Deutsch Collection/Corbis; 50TRC Hovercraft Museum Trust; 51BRC Bettmann/Corbis; 51C © Hovercraft Museum Trust; 54CLC Bettmann/Corbis; 54TR Science Photo Library; 55BLC Roger Ressmeyer/Corbis; 55CRC Roger Ressmeyer/Corbis; 56TR Science Museum/Science & Society Photo Library; 56CL Mary Evans Picture Library; 56–57BC Harry Ransom Humanities Research Centre, The University of Texas at Austin; 57TR Courtesy of Richard Platt; 57C Science Museum/Science & Society Photo Library/Daguerre; 58BL Robert Harding Picture Library/USA - Boston Massachusetts, The Polaroid Factory; 58TR Science Museum/Science & Society Photo Library; 59CL Popperfoto; 59BR Polaroid Corporation; 59TR National Museum of Photography, Film & TV/Science & Society Picture Library; 62CR Mary Evans Picture Library; 62CL Library of Congress/Science Photo Library; 62BL Science Museum/Science & Society Photo Library; 63TR Jean-Loup Charmet/Science Photo Library; 64BL Courtesy of Richard Platt; 64TR Courtesy of Richard Platt; 65TR Mary Evans Picture Library; 66CL (head only) Science Photo Library; 67TR Courtesy Garibaldi-Meucci Museum, Staten Island, NY; 66–67BL, BR Mary Evans Picture Library; 66TR J-L Charmet/Science Photo Library; 67 (head only) Mary Evans Picture Library; 68CL Science Museum/Science & Society Photo Library; 68TR Science Museum/Science & Society Photo Library; 69CR From the collections of Henry Ford Museum & Greenfield Village; 70BL, 71 BR Science Photo Library; 70–71B Courtesy of Richard Platt; 71TR Science Museum/Science & Society Photo Library; 72C Nik Wheeler/Corbis; 72BR Bettmann/Corbis; 73BR D Roberts/Science Photo Library; 73TC Bettmann/Corbis; 74BLC Bettmann/Corbis; 74BLC Bettmann/Corbis; 74TR Science Museum/Science & Society Picture Library; 75TR Mary Evans Picture Library; 75BL Musee de Radio-France, Paris, France/Bridgeman Art Library; 78CL Courtesy of State Archives of Michigan; 78C From the collections of Henry Ford Museum & Greenfield Village Research Centre; 78–79 Minnesota Historical Society/Corbis; 80BL Science Museum, London, UK/Bridgeman Art Library; 80TR Corbis; 80BR Science Museum/Science & Society Picture Library; 81TRC Hulton-Deutsch Collection/Corbis; 81BL Science Museum/Science & Society Picture Library; 81BC Science Museum/Science & Society Picture Library; 81BR Science Museum/Science & Society Picture Library; 82TR Mary Evans Picture Library; 83CR Mary Evans Picture Library; 85BR Sony Computer Entertainment Europe/Sony Playstation®; 85TRC 2000 by Maury Markowitz, www.sympatico.ca; 86CL Sony Corporation; 86C © James A. Sugar/Corbis; 87BRC Bettmann/Corbis; 88C www.geocities.com/Heartland/valley.htm; 88C Henry Horenstein/Corbis; 88BLC © Telehouse; 88TR Charles E. Rotkin/Corbis

The publisher would like to thank the following illustrators:
Mark Bristow 2–3, 12–13C, 21TR, 29BL, 30C, 44BL, 44–45C, 46BL, 54BR, 55TR, 62–63C, 82–83BC, 87TR, 86–87C; Mike Buckley 36–37C; Tom Connell 14–15R, 22–23C, 34–35TL; Richard Platt 17TR, 46BR, 48TR; Jurgen Ziewe 8–9C, 26BL, 64–65C, 68–69C, 84–85C

The author Richard Platt would like to thank the following for their assistance:
John Gustafson; Norman Krim, Raytheon archivist; Lori Lohmeyer, Nation's Restaurant News; Paul Stevenson, Stockton Reference Library; Ross Macmillan and Hugh Turral, Agricultural Engineering, University of Melbourne; Marilyn Ward, Royal Agricultural and Horticultural Society of South Australia